T0184782

Security Analytics for the Internet of Everything

Security Analytics for the Internet of Everything

Edited by
Mohiuddin Ahmed, Abu S. S. M. Barkat Ullah,
and Al-Sakib Khan Pathan

CRC Press
Taylor & Francis Group
Boca Raton London New York

CRC Press is an imprint of the
Taylor & Francis Group, an **informa** business

CRC Press
Taylor & Francis Group
6000 Broken Sound Parkway NW, Suite 300
Boca Raton, FL 33487-2742

First issued in paperback 2022

© 2020 by Taylor & Francis Group, LLC
CRC Press is an imprint of Taylor & Francis Group, an Informa business

No claim to original U.S. Government works

ISBN 13: 978-1-03-240069-3 (pbk)
ISBN 13: 978-0-367-44092-3 (hbk)
ISBN 13: 978-1-003-01046-3 (ebk)

DOI: 10.1201/9781003010463

**Visit the Taylor & Francis Web site at
http://www.taylorandfrancis.com**

**and the CRC Press Web site at
http://www.crcpress.com**

Dedicated to

"My Loving Wife: Raiyan"

Mohiuddin Ahmed

"My Loving Wife, Jhumu, & Dearest Son, Reyan"

Abu S. S. M. Barkat Ullah

"My Two Little Daughters: Rumaysa and Rufaida"

Al-Sakib Khan Pathan

Contents

SECTION IV   CYBER CHALLENGES

Preface

Introduction

The Internet is making our daily life as digital as possible, and this new era is called Internet of Everything (IoE). While it creates new opportunities, it also presents a fundamental challenge related to the security of the environment. To establish secure connections among people, processes, data, and things, security needs to be ubiquitous. The physical and cyber security must work collaboratively to protect the networks, applications, devices, data, and users, which are the building blocks of IoE. Due to the increase in connected devices, big data, and automation, system data need to be processed as securely as possible. The increased connectivity will create more room for cyber criminals to exploit the system's vulnerabilities. In fact, even in today's dynamic threat environment, the role of security is evolving, becoming increasingly complex, and considered an especially critical factor for any successful business. According to Cisco, 65% of companies could not prevent a breach even with the use of state-of-the-art prevention mechanisms, while 33% organizations took more than two years to identify a breach! As expected, around 52% of the organizations lost reputation due to a breach. Therefore, to secure the IoE, it is imperative to use a holistic approach. There is a clear need for collaborative work throughout the entire value chain of the network. In this context, this book will address the Security Analytics for the Internet of Everything, which will provide a big picture on the concepts, techniques, applications, and open research directions in this area. In addition, the book is expected to serve as a single source of reference for acquiring the knowledge on the technology, process, and people involved in the next-generation cyber security.

Objective of the Book

This book is about compiling the latest trends and issues about emerging technologies, concepts, and applications which are based on cyber security. It is written for graduate students in universities, researchers, academics, and industry practitioners working in the areas of cyber security, data science, and machine learning.

Target Audience and Content

The target audience of this book comprises graduate students, professionals, and researchers working in the field of cyber security, computer networks, communications, and Internet of Everything (IoE). This book includes chapters covering the fundamental concepts, relevant techniques, and interesting challenges. The chapters are categorized into four groups with a total of 10 chapters. These chapters have been contributed by authors from different countries across the globe.

Section I: Emerging Security Trends
Section II: Cyber Governance
Section III: Artificial Intelligence in Cyber Security
Section IV: Cyber Challenges

The first section contains three chapters that cover the emerging trends of cyber security. These chapters reflect important knowledge areas such as blockchain and dark web. The chapters highlight the important roles blockchain is playing in today's networked environment. These topics are considered as emerging trends for security analytics. The second section has two chapters reflecting cyber governance issues such as public sector issues and a project manager's view on cyber security initiatives. The third section is dedicated to the techniques supported by artificial intelligence to address different cyber security issues and performance analysis. The last section includes two chapters on cyber challenges covering basic issues such as cryptographic attacks and challenges faced by military organizations.

Acknowledgments

We are grateful to the Almighty Allah for blessing us with the opportunity to work on this book. It is another incredible book-editing experience, and our sincere gratitude is to the publisher for facilitating the process. This book-editing journey enhanced our patience, communication, and tenacity. We are thankful to all the contributors, critics, and the publishing team. Last but not the least, our very best wishes are for our family members whose support and encouragement contributed significantly to complete this book.

Mohiuddin Ahmed, PhD
Edith Cowan University, Australia

Abu S. S. M. Barkat Ullah, PhD
Canberra Institute of Technology, Australia

Al-Sakib Khan Pathan, PhD
Southeast University, Bangladesh

Editors

Mohiuddin Ahmed attained his PhD in Computer Science from the University of New South Wales (UNSW Australia). He has made practical and theoretical contributions in big data analytics (summarization) for a number of application domains and his research has a high impact on data analytics, critical infrastructure protection (IoT, smart grids), information security against DoS attacks, false data injection attacks, and digital health. He is currently working as a Lecturer in Computing and Security Sciences in the School of Science at Edith Cowan University (ECU), Australia. Prior to joining ECU, he served as a Lecturer in the Centre for Cyber Security and Games at Canberra Institute of Technology (CIT) and was also involved with CIT's Data Strategy Working Group. He is currently exploring blockchain for ensuring security of healthcare devices, securing the prestigious ECU Early Career Researcher Grant. Mohiuddin has led edited books on Data Analytics, Cyber Security and Blockchain. Previously, he has worked in the areas of text mining and predictive analytics in the artificial intelligence division at MIMOS, Malaysia. Currently, Mohiuddin is an editorial advisory board member of Cambridge Scholars Publishing Group in the UK and Associate Editor of the *International Journal of Computers and Applications* (Taylor & Francis Group).

Abu S. S. M. Barkat Ullah is working as Head of the Centre for Cyber Security and Games at Canberra Institute of Technology. He attained his PhD from the University of New South Wales (UNSW), Australia, with significant contribution in the areas of computational intelligence, genetic computing, and optimization. He is currently the principal investigator of a healthcare cyber security research project. He has published in reputed venues of computer science and is actively involved in both academia and industry. He holds a Bachelor of Science Degree in Computer Science with outstanding results and achieved a Gold Medal for such performance. He is also one of the pioneers in introducing false data injection attacks in the healthcare domain.

Al-Sakib Khan Pathan received his PhD degree in Computer Engineering in 2009 from Kyung Hee University, South Korea, and B.Sc. degree in Computer Science and Information Technology from Islamic University of Technology (IUT), Bangladesh, in 2003. He is currently an Associate Professor at the

Computer Science and Engineering (CSE) department, Southeast University, Bangladesh. Previously, he was with the Computer Science department at the International Islamic University Malaysia (IIUM), during 2010–2015, and with the CSE department at BRAC University, Bangladesh, during 2009–2010. He also worked as a Researcher at Networking Lab, Kyung Hee University, South Korea, from September 2005 to August 2009, where he completed his MS leading to PhD. His research interests include wireless sensor networks, network security, cloud computing, and e-services technologies. Currently, he is also working on some multidisciplinary issues. He is a recipient of several awards/best paper awards and has several notable publications in these areas. He has served as a Chair, Organizing Committee Member, and Technical Program Committee (TPC) member in numerous international conferences/workshops, including top tier and prestigious events like INFOCOM, GLOBECOM, ICC, LCN, GreenCom, AINA, WCNC, HPCS, ICA3PP, IWCMC, VTC, HPCC. He was awarded the IEEE Outstanding Leadership Award and Certificate of Appreciation for his role in IEEE GreenCom'13 conference. He is currently serving in various editorial positions like Series Editor of ICT Book series by CRC Press/Taylor & Francis Group; Deputy Editor-in-Chief of *International Journal of Computers and Applications*, Taylor & Francis Group; Associate Technical Editor of *IEEE Communications Magazine*; Editor of *Ad Hoc and Sensor Wireless Networks*, Old City Publishing, and *International Journal of Sensor Networks*, Inderscience Publishers; Associate Editor of *International Journal of Computational Science and Engineering*, Inderscience; Area Editor of *International Journal of Communication Networks and Information Security*; Guest Editor of many special issues of top ranked journals; and Editor/Author of 15 published books. One of his books has been included twice in the Intel Corporation's Recommended Reading List for Developers, 2nd half 2013 and 1st half of 2014; 3 books were included in IEEE Communications Society's (IEEE ComSoc) Best Readings in Communications and Information Systems Security, 2013; 2 other books were indexed with all the titles (chapters) in Elsevier's acclaimed abstract and citation database, Scopus, in February 2015; and a seventh book is translated to simplified Chinese language from English version. Also, 2 of his journal papers and 1 conference paper were included under different categories in IEEE Communications Society's (IEEE ComSoc) Best Readings Topics on Communications and Information Systems Security, 2013. He also serves as a referee of many prestigious journals. He has received some awards for his reviewing activities like one of the most active reviewers of IAJIT three times, in 2012, 2014, and 2015; Outstanding Reviewer of Elsevier Computer Networks (July 2015); and Elsevier JNCA (November 2015), just to mention a few. He is a Senior Member of the Institute of Electrical and Electronics Engineers (IEEE), USA.

Contributors

Tarem Ahmed
Department of Computer Science and
  Engineering
Independent University Bangladesh
  (IUB)
Dhaka, Bangladesh

Hazaa Al Fahdi
Academic Centre of Cyber Security
  Excellence, School of Science
Edith Cowan University
Perth, Australia

Zubair Baig
School of Information Technology
Deakin University
Melbourne, Australia

Pushpita Chatterjee
Department of Modeling, Simulation
  and Visualization Engineering
Old Dominion University
Norfolk, Virginia

Khalid Chauhan
DIG, Police
Government of Azad Jammu and
  Kashmir
New Mirpur City, Pakistan

Kevin Chong
Academic Centre of Cyber Security
  Excellence, School of Science
Edith Cowan University
Perth, Australia

Salva Daneshgadeh
Department of Information Systems
Middle East Technical University
Ankara, Turkey

Raja Datta
Department of Electronics and
  Electrical Communication
  Engineering
IIT Kharagpur
Kharagpur, India

Uttam Ghosh
Department of Electrical Engineering
  and Computer Science
Vanderbilt University
Nashville, Tennessee

Peter Hannay
School of Science
Edith Cowan University
Perth, Australia

Bushra Hossain
Information Communication
    Engineering (ICE) Department
Bangladesh University of
    Professionals
Dhaka, Bangladesh

Ahmed Ibrahim
Academic Centre of Cyber Security
    Excellence, School of Science
Edith Cowan University
Perth, Australia

Syed Mohammed Shamsul Islam
Academic Centre of Cyber Security
    Excellence, School of Science
Edith Cowan University
Perth, Australia

Muhammad Imran Malik
Academic Centre of Cyber Security
    Excellence, School of Science
Edith Cowan University
Perth, Australia

Ian Noel McAteer
Academic Centre of Cyber Security
    Excellence, School of Science
Edith Cowan University
Perth, Australia

Kazım Rıfat Özyılmaz
Department of Computer Engineering
Bogazici University
İstanbul, Turkey

Tahmina Rashid
Faculty of Arts & Design
University of Canberra
Canberra, Australia

Danda B. Rawat
Department of Electrical Engineering &
    Computer Science
Howard University
Washington, DC

Shahrin Sadik
Department of Computer Science and
    Engineering
International Islamic University of
    Chittagong
Chittagong, Bangladesh

Munir A. Saeed
School of Engineering and Information
    Technology
UNSW Canberra
Canberra, Australia

Arda Yurdakul
Department of Computer Engineering
Bogazici University
İstanbul, Turkey

Guanglou Zheng
Security Research Institute
Edith Cowan University
Perth, Australia

EMERGING
SECURITY TRENDS

Chapter 1

Critical Analysis of Blockchain for Internet of Everything

Bushra Hossain

Bangladesh University of Professionals

Contents

1.1 Introduction

The Internet of Things (IoT) is a massive network of connected things, which allows them to interact and share data, and perform automated functions to support human users. "Things" include almost any electronic machine with an "On" and "Off" switch (1 and 0) and incorporate a connection to the internet and with each other. The term IoT was first introduced by K. Ashton in 1999 [1]. However, the concept of IoT came into limelight after a document released in 2015 by IEEE IoT Initiative [2]. It describes the technology used for implementing IoT along with the architecture and applications in different domains. Since then, it has become an intense research topic for the people from the tech space. Gartner's research discloses that the number of connected IoT devices is expected to grow to 20 billion by 2020 [3]. The influencing factors that determine the steady growth of IoT devices are: (1) the expansion of broadband internet, (2) decreasing cost of internet connection, (3) manufacturing of WiFi based devices, (4) using built-in sensors in devices, and (5) falling prices of embedded computers and smartphones, and so on.

With digitization in every sector of human life, IoT is expanding every day through the inclusion of new devices. Moreover, the availability of the internet has further enhanced the process. For instance, it is expanding from personal healthcare to agriculture, and from housing to smart city. These devices lead to a pervasive collection of data that contains personal as well as public information. Access to this valuable information may create opportunities for cyber-crime. Therefore, security is a major concern for emerging IoT industries. Furthermore, the heavily centralized IoT infrastructures require all devices to be connected and authenticated through the server and create opportunities to a single point of failure. Thus, scalability and security are the main challenges in adopting IoT in the real world due to the excessive number of connected devices in a centralized network. Therefore, transforming the IoT into the decentralized approach offers a time-worthy solution to the single point of failure and facilitates the long-term growth of IoT.

Establishing a peer-to-peer communication can help in achieving the decentralized approach for IoT as mentioned above. However, this approach alone cannot assure the solution to security issues that arise in IoT. The blockchain is the latest integration to the field of IoT with the intention to achieve a higher level of autonomous security [4]. The motivation behind the integration of blockchain in the field of IoT is to allow devices to function autonomously and securely using their "distributed ledger" and "core cryptography" technologies. In addition, a decentralized blockchain network has the ability to handle billions of transactions among the IoT devices, thereby reducing the cost associated with installation and maintenance of large centralized data centers. Furthermore, the decentralization model of blockchain also mitigates the risk of a single point of failure, which is allied with the centralized IoT architecture. In these circumstances, an extensive study and research are required to leverage blockchain in an IoT network. This chapter covers the integration of blockchain with IoT including proposed architectural layouts, benefits of integration, and the implementation challenges.

1.1.1 Chapter Roadmap

The rest of the chapter is organized as follows. Section 1.2 covers the overview of IoT including its key features, architecture, and challenges. Section 1.3 discusses the importance of the integration of blockchain with IoT. Section 1.4 contains the existing challenges of IoT–blockchain combination with some potential solutions. Finally, Section 1.5 concludes the chapter.

1.2 Overview of IoT

IoT enables the communication of different devices over the internet to share their data. It provides deeper automation, analysis, and integration within a system to achieve more comfort and ease to the user. It enables connectivity of those devices that have never been connected and performed like smart devices such as air-cooler, refrigerator, car, etc. Thereby, there are myriad benefits of incorporating IoT in our lives.

1.2.1 Key Features of IoT

IoT is not limited to a few features only, rather the ongoing expansion of IoT extended its features infinitely these days. The most basic features are described below to understand IoT functions more clearly [5,6].

- **Intelligence**: Intelligence is one of the key features of IoT devices. Intelligence can be something as simple as setting the temperature of the air-cooler by knowing the outside weather. IoT transforms virtually anything to smart or intelligent using its capabilities of collecting data from the ambiance and applying artificial intelligence algorithms within a network.

- **Connectivity**: IoT is the network of connected things. Connectivity is necessary to share the sensor data. IoT devices connect to the edge devices to process data faster and then the gateways send data to the cloud for analysis. This requires huge scalability in the network space to connect billions of smart devices. However, IoT opens up a new dimension in networking where networks are no longer exclusively tied to major providers and can exist on a much smaller and cheaper scale.

- **Sensors**: Using sensors in devices are the most common features of IoT. The level of intelligence depends on the effectiveness of these sensors in collecting data. The distinction of IoT is based on the varieties of this feature.

- **Things-related services**: Basically, IoT is offering things-related services. Everything is included in the "things" domain. Therefore, this new technology will change both the physical world and the information world simultaneously.

- **Heterogeneity**: IoT has vast applications in diversified fields. Thereby, several different systems, architecture, protocols, and standards are used by IoT devices. Therefore, it requires the interaction of different hardware platforms and networks with each other. This makes the IoT heterogeneous in nature.

- **Dynamic nature**: IoT devices deal with different types of data from the environment. The changes in their environment are collected as a form of data. By analyzing those data, IoT devices can change their state, e.g., from on to off, waking up from sleeping, connected and/or disconnected, etc. Therefore, dynamic in nature is another prime feature of IoT devices.

- **Enormous scale**: The increasing rate of connecting new devices in IoT networks is incredible. A report stated that 5.5 million new devices are adding in the IoT networks every day [3]. Even the management of a plethora of data generated from these devices is also critical.

- **Safety**: Ensuring the safety of devices is one of the key features of IoT. IoT devices generate and share data which makes them vulnerable to data manipulation.

1.2.2 Layered Architecture of IoT

There are various architectures exist for IoT proposed by different researchers and organizations [7,8–14]. The five-layer architecture is the most common among them [7–9,13–82]. A detailed discussion of the five-layered architecture (Figure 1.1) of IoT is given as follows:

- **Device/Perception/Object/Sensor Layer**: This is the bottom layer of IoT architecture. It consists of data collection devices and different types of sensors and actuators. They perform different actions such as querying and sensing location, humidity, temperature, motion, acceleration, vibration, heat, weight, chemical changes, etc. to collect data from the environment.

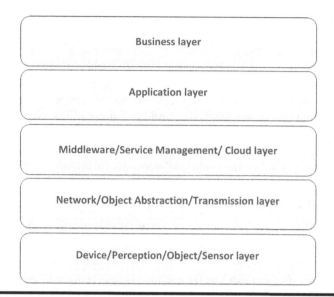

Figure 1.1   The layered architecture of IoT.

Data acquisition and processing are the main critical parts of this layer. Finally, the digitized data are transferred securely to the next layer.

■ **Network/Object Abstraction/Transmission Layer**: Networking capabilities and transport capabilities are the two main functions provided by the network layer. This layer transfers data from the device layer to the middleware layer. Various technologies are used to transfer data from source to destination like GSM, 3G, RFID, UMTS, Bluetooth, WiFi, infrared, Zigbee, etc.

■ **Middleware/Service Management/Cloud Layer**: The middleware layer has two capabilities. One is generic support capabilities, and the other is specific support capabilities. Generic support capabilities are common capabilities of IoT applications, such as data processing, data storage, etc. Specific support capabilities are particular capabilities for diversified IoT applications, such as ubiquitous computation, making decision, etc.

■ **Application Layer**: The application layer provides the requested services to the customer. It also provides an interface to the business layer. Some of IoT applications are smart home, smart city, automated car, smart agriculture, smart healthcare, intelligent transport system, etc.

■ **Business Layer**: The success of IoT applications depends on the business models built in this particular layer. The main task of this layer is to manage all the services and applications. Besides business models, it also generates flow charts and graphs based on the results received from the application layer. The future strategies and actions are determined from the outcome of this layer.

1.2.3 Difference between the Traditional Networks and IoT

The notable difference between the traditional networks and IoT is the availability of the amounts of resources at the end devices [66]. To highlight, these differences are necessary for the development of security and privacy of IoT systems. Primarily, IoT devices are the resource-constrained embedded devices that features lack of memory, shortage in computing power, small size, and low battery status, etc. On the other side, devices in the traditional networks comprise ample resources. Powerful computers, smartphones, and servers are examples of the traditional networks. The prime differences between the IoT and traditional networks are highlighted in Table 1.1.

1.2.4 Issues in Centralized IoT Model

Current IoT architectures are based on a centralized client/server model [16]. In this model, all devices need to authenticate and communicate through the central server. The current centralized architecture is used to support only the small-scale IoT networks. In the client/server model, server resources such as computing power, memory, storage, etc. need to scale up with the increasing number of clients connected.

Table 1.1 Differences between the Traditional Networks and IoT

Traditional Networks	*IoT*
Traditional networks can use complex and multi-factor security protocols without considering the resource constraint	IoT demands a lightweight algorithm to minimize the resource constraint while ensuring security
Traditional networks are connected through media such as fiber optics, WiFi, DSL/ADSL, 4G, and LTE which facilitate speed and security	Communications among the IoT devices are implemented through end devices such as 802.11a/b/g/n/p, 802.15.4, LoRa, NB-IoT, Zigbee, and Sig-Fox which lack bandwidth and power
Devices in the traditional networks comprise similar configuration, protocols, and data formats	IoT devices are application-specific and heterogeneous. Therefore, they contain different configuration, protocols, data contents, and formats
Traditional networks use multi-level security gyre with the combination of network perimeter defense (e.g., firewalls and IDS/IPS) and host-based defense mechanisms (e.g., antivirus and security/software patches)	The absence of host-based defense mechanisms and lack of updating security/software patches make IoT devices more vulnerable to the security threats. Employing only the conventional perimeter defense mechanism cannot provide robustness in security to the IoT devices

It is reported by Cisco that the number of IoT devices will reach 20 billion by 2020 [83]. Therefore, managing the huge expansion of the IoT by the centralized architecture in the future will be challenging [15]. In addition, it will also increase the cost of handling a huge amount of communication among IoT devices. The centralized architecture is also vulnerable to the single point of failure which may cause disruption of the whole network [17,22]. Existing centralized models provide privacy, security, and data-handling mechanisms trusting through a third-party entity. Trust on the third-party entities has made the centralized model susceptible to data manipulation and cyber-attacks [21].

Aforementioned facts drive centralized IoT architecture to a decentralized one. Presently, blockchain is considered to be a popular decentralized technique for IoT architectural design . In this context, the integration of blockchain with IoT offers a new paradigm for managing expansive IoT devices.

1.3 Blockchain in IoT: Future of the Digital World

The recent trend of integrating different technologies to achieve their best performance is highly appreciable. Doing so, sometimes the integration process evolves new technologies with greater efficiency or benefits to the existing technologies by eliminating their drawbacks. In this context, the integration of blockchain with IoT has the same potential to bring a revolution in the information technology (IT) industry as the internet. The unique features of blockchain can complement IoT by developing a secure distributed technology in the future. Based on these features, some of the potential benefits and motivations for adopting blockchain in IoT are discussed below (Figure 1.2) [16,18–19]:

- **Resilience**: Data integrity is highly preferable in IoT applications. Therefore, the resilience of any deterioration and leakage to data is essential in IoT frameworks. In blockchain, each peer carries a copy of the transaction records, which helps to maintain data integrity and keeps the IoT framework more resilient.
- **Identity**: Identity management in IoT involves identification of devices, sensors, monitors, users, and authorization of their access to data. Traditional identity management approaches in IoT have a shortfall of reliance and trust on third-party authorization. The rise for blockchain-based decentralized identity solutions can provide a distributed and trustworthy authentication and authorization process for IoT applications.
- **Adaptability**: One of the challenges faced by traditional centralized IoT network is that it limits the interoperability because of the heterogeneous IoT devices. It is established that blockchain performs well in heterogeneous hardware platforms. Thus, using blockchain can add a higher degree of adaptability in IoT devices.
- **Security and privacy**: The massive growth and distributed nature of IoT networks open up doors for security and privacy challenges. Blockchain

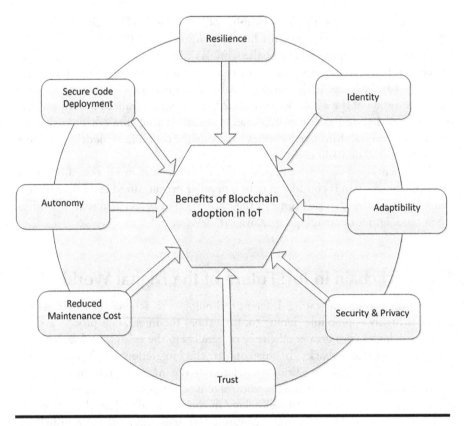

Figure 1.2   Benefits of blockchain adoption in IoT.

inherits many significant qualities like records immutability, distributed consensus, reliance in a trustless environment, and so on. These characteristics of blockchain will facilitate achieving decentralized security and privacy in IoT devices.

■ **Trust**: Traditionally, IoT services are based on centralized infrastructure where trusting a third party is a requirement. But, the "trustless" nature of blockchain removes this need of trust in centralized authorities to handle the IoT data, resulting in an improvement in data integrity by preventing any security threats from a malicious third party .

■ **Reduced maintenance costs**: The deployment and operation costs of IoT are huge as the present security protocols are based on sophisticated cryptographic computations. Blockchains have the potential to significantly reduce maintenance cost as they do not need intermediary servers. For instance, Sia4 is a blockchain-based, simple, decentralized storage in which the available storage can be given as rent instead of using dedicated servers to utilize the available space and reduce cost.

- **Autonomy**: In blockchain framework, devices interact with each other without any third-party involvement. Therefore, it empowers the features of next generations like autonomy, data monetization, etc.
- **Secure code deployment**: The secure and immutable storage of blockchain ensures code safety and security while deploying into IoT devices.

1.3.1 Blockchain-Based IoT Architecture

From the literature review, we identified four types of integration schemes in blockchain-based IoT architecture [16,19]. While forming a blockchain-based IoT network, the underlying architecture is determined by the communication location (i.e. inside IoT, through IoT gateway, through blockchain or a hybrid design of IoT and blockchain, etc.) of IoT devices. The following sections broadly discuss different integration schemes of blockchain and IoT (Figure 1.3).

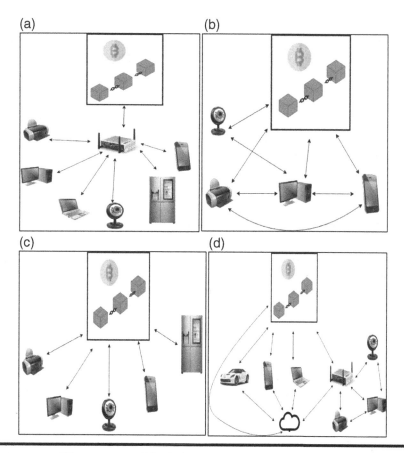

Figure 1.3   Different types of blockchain-based IoT architecture: (a) IoT gateway–blockchain, (b) IoT–IoT, (c) IoT–blockchain, and (d) hybrid.

1.3.1.1 IoT Gateway–Blockchain

In this approach, end-points of a blockchain network are the IoT gateways. The IoT devices are registered to the gateways, and the gateway provides transactions to the blockchain, as shown in Figure 1.3a. It ensures the traceability as all communications with blockchain is performed through the specific IoT gateway. In this approach, P2P technologies like IPFS 9 and BitTorrent are used for data transfer, which minimizes the requirements to store all of the transaction records on the blockchain. As the IoT devices do not have a direct transaction with blockchain, this integration scheme does not ensure a detailed decentralized network. This integration scheme of IoT with blockchain would be useful in scenarios where authentication of communications between the connected devices in separate blockchain-enabled gateways is required.

1.3.1.2 IoT–IoT

In this approach, end-points of a blockchain network are interconnected edge devices. Since the devices are able to communicate with each other offline, it reduces latency and enhances security. Instead of using blockchain for interactions among the IoT devices, a part of IoT data is stored in the blockchain. It requires routing and discovery protocols to communicate with the IoT devices. This limited edition integration scheme of IoT with blockchain would be suitable in the scenarios where reliable IoT data are available for the IoT interactions with high throughput and low latency.

1.3.1.3 IoT–Blockchain

In this approach, all the transactions are performed through blockchain. Therefore, it ensures the traceability of all communications as they are recorded. Consequently, recording all the transactions in blockchain increases bandwidth and storage requirements, which is a key research challenge of blockchain-IoT integration. This integration scheme of IoT with blockchain would be suitable in the scenarios where a large number of autonomous IoT devices are required.

1.3.1.4 Hybrid

In the hybrid approach, IoT users can make decisions which transactions should go through blockchain and which transactions should be directly shared between the IoT devices. One of the primary challenges in hybrid approach is faced while choosing the right transaction mode in run time. The hybrid approach combines all other integration schemes to get benefits of decentralization through blockchain and real-time IoT interactions through directly shared IoT devices. Figure 1.3d illustrates the hybrid approach of blockchain–IOT integration scheme. This integration scheme of IoT with blockchain would be suitable in the scenarios where a balanced, blockchain-based IoT network is required. A fog computing can also be used to facilitate the hybrid scheme of a blockchain-based IoT network.

1.3.2 Security Issues

A secured system needs to satisfy three main requirements: Confidentiality, Integrity, and Availability known as CIA [24,25]. Confidentiality ensures that information is denied to unauthorized individuals, entities, or processes. Integrity assures that data cannot be modified in an unauthorized or undetected manner. Availability guarantees that each service or data is available to the user when it is needed. In this section, the security issues faced by centralized IoT infrastructures are discussed followed by highlighting the potential blockchain solutions from the literature that helped to improve the CIA in IoT.

1.3.2.1 Security Issues in Centralized IoT Model

In October 2016, Dyn—a US-based popular DNS provider—was crippled by DDoS (distributed denial-of-service) cyber-attacks [23,26]. An investigation by Dyn reveals that these attacks originated from "tens of millions of IP addresses," including some IoT devices like webcams, baby monitors, home routers, and digital video recorders [28,29]. These IoT devices had been compromised by a malware called Mirai, which creates a robot network, or botnet, and sends millions of fake messages to victims' computer systems to launch DDoS [27]. As per the security experts' opinion, phishing emails first infect a computer or a home network, and then it spreads to other devices, such as DVRs, printers, routers, and internet-connected cameras employed by stores and businesses for surveillance [30].

The aforementioned incident has clearly pointed out the presence of huge security risks at the current IoT network. At present, most of the IoT frameworks are based on the centralized client–server model which creates the essential security challenges to the IoT. There are significant security challenges of centralized IoT model which is listed below:

1. Each node in an IoT network is prone to fail on cyber-attack such as DDoS [32].
2. A set of corrupted nodes and devices in the IoT edge can function simultaneously to collapse the service, as seen in recent botnet attacks [33].
3. Centralized IoT framework is vulnerable to the central point of failure which is a threat to confidentiality, availability, and authorization [22].
4. Centralized IoT does not provide in-built assurance of integrity and confidentiality with IoT data as it involves third-party security service. For instance, confidentiality attacks arise from identity spoofing and analyzing routing and traffic information, etc. [19]

5. IoT data integrity is challenged by Byzantine routing information attacks and modification attacks [31]. In centralized IoT configuration, data integrity is also a threat to injection attacks which affect decision-making applications by injecting false measures in data.

1.3.2.2 Potential Blockchain Solutions from Literature

A plethora of solutions have been proposed for improving security of the IoT using blockchain. Some potential solutions are enumerated in Table 1.2.

Table 1.2 Blockchain-Inspired Potential Security Solutions for Internet of Things

Fundamental Security Goal	Authors	Proposed Solution For	Method Used
Confidentiality	Aitzan et al. [67]	Confidentiality of energy trading smart grid	Distribution system operators (DSOs) to manage blockchain addressing and security
	Ali et al. [68]	Facilitating secure IoT communications	Multi-tier solution using blockchain and IPFS off-chain
	Alphand et al. [69]	Securing authorization access to IoT devices	Blockchain's smart contract security policy and OSCAR security model
	Axon et al. [70]	Solving existing PKI's flaws	Blockchain-based PKI
	Boudguiga et al. [78]	Securing deployment of updates for IoT devices	Storing software updates in the blockchain for IoT devices
	Cha et al. [71]	Securing BLE-based IoT devices	Blockchain-connected gateways are used
	Dorri et al. [20]	Ensuring privacy and security of an IoT environment	A multi-tier blockchain architecture, Diffie–Hellman encryption algorithm

(Continued)

Table 1.2 (*Continued*) Blockchain Inspired Potential Security Solutions for Internet of Things

Fundamental Security Goal	*Authors*	*Proposed Solution For*	*Method Used*
Confidentiality (continued)	Fair access [74]	Maintaining fair-access control in an IoT environment	Access privileges granted with signed cryptocurrency transactions
Integrity	Ali et al. [68]	Facilitating secure IoT communications	Shared keys are used in blockchain transactions for IPFS content addressing
	Biswas et al. [75]	Securing smart cities	Tamper-proof blockchain records
	Dorri et al. [20]	Ensuring privacy and security of an IoT environment	Multi-tiered blockchain architecture maintains records of cloud-based data
	Zyskind et al. [72], Shafagh et al. [73]	A secure distributed platform for computing and storing IoT data	Tamper-proof blockchain records of data stored in DHT
	Liu et al. [76]	Ensuring data integrity for cloud-based IoT	Blockchain-based querying framework for data integrity verification
	Lee et al. [77]	Securely update firmware for embedded devices in IoT	Blockchain transactions for firmware updates to embedded devices
	Boudguiga et al. [78]	Securing deployment of updates for IoT devices	Storing software updates in the blockchain for IoT devices
	Steger et al. [79]	Securing software updates available to smart vehicles	Multilayered blockchain architecture is used for propagating software update

(Continued)

Table 1.2 (*Continued*)   Blockchain Inspired Potential Security Solutions for Internet of Things

Fundamental Security Goal	Authors	Proposed Solution For	Method Used
Integrity (continued)	Yang et al. [80]	A reputation system for assessing credibility of incoming messages in vehicular networks	Blockchain-based message rating system is used
Availability	Alphand et al. [69]	Securing authorization access to IoT devices	A fault-tolerance system using blockchain for IoT authorization model
	Ali et al. [68]	Facilitating secure IoT communications	Transactions for IPFS file access over public Ethernet blockchain
	Chakraborty et al. [8]	Optimizing computational load in IoT network	A combination of traditional centralize structure with distributive multilay-ered blockchain network is used
	Boudguiga et al. [78]	Securing deployment of updates for IoT devices	Storing software updates in the blockchain for IoT devices
	Bahga et al. [81]	Developing a secured smart manufacturing system	Blockchain platform for industrial Internet of Things (BPIIoT), a cloud-based blockchain technology is used

1.3.3 Establishing Security and Privacy through Blockchain: A Case Study of Smart Home

A smart home is an integrated home automation system to maximize the comfort, peace, and safety of the people living there [34]. In a smart home, the owner can remotely control, activate, and deactivate devices, as well as the devices can be controlled automatically. This system transforms the home to an intelligent entity to adopt the owner's preferences and habits. Doing so, connected devices in a smart home generate data and communicate with each other to extract information that can reveal the owner's living characteristics. The smart devices inside the home may exchange data within themselves as well as with the external entities to provide certain services as per the owner's preferences. For instance, the proximity sensor sends data to the switch to turn off the lights automatically when someone leaves the room. Therefore, a potential risk of data leakage is associated in such systems while communicating among the smart entities for data exchange as well as during storage of data.

Dorri et al. [20] designed a blockchain-based multi-tier architecture for an IoT environment which meets the fundamental security goals of confidentiality, integrity, and availability. The design consists of three core tiers: smart home, overlay network, and cloud storage. The smart home tier constitutes smart devices and a single miner with local storage. The home miner is a device that primarily manages a local blockchain. The miner centrally processes all incoming and outgoing transactions of the smart home. It collects all the transaction into a block and appends a full block to the blockchain. In addition, the miner also manages transaction access policies for ensuring security and local storage for providing additional capacity. The miner creates a block and registers it to the blockchain whenever a device is added to the smart home.

Each block of the blockchain contains two headers: block header and policy header along with transactions information. The block header holds the link to the previous block, and the policy header stores access permission of transactions in the blockchain. There are different transactions available in the blockchain-based smart home such as store transaction is to store data by devices, access transaction is to access cloud storage by the service provider or homeowner, monitor transaction is to monitor devices information by service provider or homeowner, and genesis transaction is to add a new device and remove transaction is to remove a device. The aforementioned transactions secure their communications using a shared key. The miner manages the distribution and creation of these keys using the generalized Diffie–Hellman algorithm.

1.4 Summary on Blockchain–IoT Integration

IoT is one of the most promising information and communication technologies in recent years. As IoT is proliferating, many researches disclose that an integration of the blockchain with IoT could prove to be vital. Though blockchain was designed for securing monetary transactions on the internet through powerful computers, it has the potential to contribute to other sectors. This section summarizes the current challenges of incorporating blockchain-based IoT and discusses some of the opportunities to work within the limitations of blockchain technology. Finally, the scope of future research is also given to enhance further development in this field.

1.4.1 Challenges and Opportunities

Though the integration of blockchain with IoT is very beneficial, blockchain technology does have limitations. Therefore, incorporating blockchain into the IoT is challenging. Some of the challenges are discussed in this section with the opportunities to overcome them.

1.4.1.1 Storage Capacity and Scalability

Storage capacity and scalability are the inherent limitations of blockchain. Every ten minutes, the bitcoin chain is growing at a rate of 1 MB per block. Considering this ten-minute time to add a new block to the chain, bitcoin's blockchain can manage a maximum of seven transactions per second [36]. Every time the updated chain is also stored among the nodes in the blockchain network. Consequently, the increasing size of nodes demands computational resources aggressively, which reduces the system's capacity scale. These inherent limitations of blockchain add much greater challenge to the IoT platform. As we know, IoT devices generate gigabytes (GB) of data in real time, which is unsuitable for blockchain to store such a large amount of data. Moreover, a blockchain-based IoT network may require to manage a large number of transactions per time unit. This is a potential bottleneck for certain blockchain networks. Though the deployment of blockchain in IoT faces those challenges, there are still some opportunities to overcome them. As we know, a limited part of the stored IoT data is useful to extract knowledge and generate actions based on inferred knowledge. Therefore, different techniques of data compression and normalization have been proposed to reduce data size to overcome storage capacity challenges. Lightweight IoT node is another concept used to avoid storage issues of blockchain, where blockchain is used only to perform transactions, but need not store it [37]. A "mini-blockchain" concept is another alternative, which is proposed to store the most recent transactions only [38,39]. Primarily blockchain has three functions: (1) coordinate the transaction processes in the network, (2) encapsulate proof-of-work (PoW) to secure the network, and (3) manage account balances and record the ownership of the coins. These three functions are combined into a single mechanism in the bitcoin blockchain which causes scalability issues. Therefore, in the "mini-blockchain" scheme the key concept is breaking the functions and providing an individual mechanism for each of them to achieve more storage capacity and scalability.

Bitcoin-NG is a next-generation blockchain protocol that eliminates scalability problems without changing bitcoin's open architecture and trust model [40]. In bitcoin, after every ten minutes, the system generates a backward-looking block that contains transactions of the preceding ten minutes. In contrast, Bitcoin-NG is a forward-looking protocol where in every ten minutes a leader is elected to maintain the transaction serials until the next leader is elected. Thereby, Bitcoin-Ng decoupled the blockchain structure into two types of blocks: key blocks and microblocks.

Key blocks are generated through PoW for selecting a leader, and they don't carry any transactional data. Microblocks store all transaction information, and several microblocks can follow a key block. Therefore, the block size remains so small. Splitting the block into key block and microblock reduces the block propagation overhead that contributes to high throughput. The use of key blocks reduces block construction time which helps in achieving low latency.

1.4.1.2 Security

Even though the blockchain technology is known as highly secured, its integration with IoT will face some security challenges due to the complex nature of IoT. Wireless communication, mobility, heterogeneity of devices, and a huge amount of accessible data affect IoT's security. The main security challenge of blockchain-based IoT is the reliability of the data generated by IoT devices. As we know, blockchain is an immutable ledger that ensures privacy and security in the chain. But, the problem arises when corrupted data arrive at the blockchain and remain corrupted. The malicious events are not always responsible for data corruption; events like failure of the devices, vandalism, and hostile environment are also causing data corruption. Sometimes the devices fail to work properly from the beginning which may cause corrupted data undetected. Therefore, a thorough investigation should be done before integrating IoT devices with blockchain. Resource restriction in IoT devices is another challenge for IoT–blockchain integration, which hinders the robust security mechanism. We know that IoT devices are resource restricted, whereas blockchain technology requires miners with huge computational resources. Therefore, the constraint limits to run proper cryptographic algorithms of blockchain in IoT and creates security breaches. Even their constraints also restrict essential firmware updates and make IoT vulnerable to bugs and malicious entities. To overcome the firmware updating challenge, a runtime-based upgrading and reconfiguration system can be placed in IoT devices. GUITAR [41] and REMOWARE [42] are such initiatives to facilitate runtime updating of firmware in IoT devices and ensure security to the blockchain-based IoT. The Filament [43] is one of the noticeable projects that provide blockchain-based secure hardware and software solutions for the enterprise and industrial IoT.

1.4.1.3 Privacy

There are two main aspects of privacy preservation in IoT: user privacy or anonymity, and data secrecy. Many applications in IoT carry confidential data; for example, healthcare IoT devices deal with confidential information of patients. Therefore, anonymity and data privacy are essential for these types of data. Though blockchain could be a solution for the privacy issue in IoT, while designing the bitcoin protocol, privacy is not enforced. Bitcoin offers transparency. In a bitcoin blockchain, all the transactions are stored and visible to the participants of the bitcoin network.

Although this level of transparency is needed to build trust in the network, such open records can be used to reveal user confidential information [44] and even their actual identity can be traced [45]. Moreover, the bitcoin blockchain does not guarantee total anonymity through a "pseudonymous" mechanism. Thereby, they are vulnerable to linking attacks also [46]. Zerocash [47] and Zerocoin [48] provide a zero-knowledge proving technique to avoid anonymity problem in bitcoin. They proposed bitcoin extensions which ensure complete anonymous transactions by hiding the sender, the receiver, and the information itself. To ensure anonymity in IoT applications, zero-knowledge proofs can be adopted to hide user identity or devices [49].

Another privacy-focused method called Monero [50] is a ring signature scheme that keeps transactions untraceable to ensure anonymity. Even the problem of data privacy of IoT devices starts from the very beginning of data collection and continues to remain at communication and application levels. Securing IoT devices from unauthorized user's access requires the integration of cryptographic software into the device. Hawk, a blockchain-based smart contract, ensures encrypted transactions on the blockchain to secure data privacy [51]. To establish data privacy and confidentiality in the private blockchain, Hyperledger-fabric uses private channels for providing encryption and sharing data [52]. Similarly, Quorum mixes cryptography and segmentation for the security of sensitive data [53]. But, these privacy-preserving techniques require a huge computational resource which is a trade-off for resource-constrained IoT devices. Therefore, further research is needed to ensure privacy-preserving computations and data analytics for adopting blockchain in IoT.

1.4.1.4 Consensus Protocol

The consensus protocol is the core part of any blockchain network. It is a common agreement on the present state of the distributed ledger among all the peers in the blockchain network. Primarily, the consensus protocol verifies the authenticity of a new block when it is added to the blockchain. This way, the consensus mechanisms achieve integrity and reliability, and establish trust between unknown peers in an untrusted distributed network. Some of the current consensus protocols are PoW [54], PoS [55], PoET [56], and IOTA [57]. They are designed for permissionless blockchains that allow equal and open rights to all participants in a blockchain network. However, among them, PoS and PoET can also be used in permissioned blockchains [58]. The common feature that is shared by all the consensus protocols as mentioned earlier is the probabilistic finality. This finality assures that all the well-formed blocks in the blockchain will not be revoked or reversed once they are committed to a blockchain. In probabilistic finality, the probability of a transaction will not be reverted increases as the block that contains the transaction goes deeper into the chain. Therefore, it results in delayed transaction confirmation. On the contrary, IoT demands immediate transaction confirmation. Therefore, these consensus protocols are not suitable for IoT. PoW is the first blockchain consensus algorithm that has worked successfully in Bitcoin. The PoW consensus algorithm forces miners to solve a computationally challenging puzzle to create a new block in the blockchain. When a miner finally solves the puzzle, the node broadcasts it to the whole network and receives a cryptocurrency prize (the incentive) provided by the PoW protocol.

A major limitation of PoW protocol is it requires huge resource consumption. To find the solution to the computationally challenging puzzle, miners consume a huge amount of computing power, which results in wastage of resources. Certain drawbacks such as high latency, low transaction rates, and high energy consumption make it unsuitable for various IoT applications. Thereby, a variety of proposals to facilitate consensus protocol have been made. To reduce high latency, an off-chain-based solution is proposed where information can be stored outside the blockchain. Though IoT has resource constraint problem, globally IoT is capable of huge computational processing. Therefore, more research should be focused on discovering the global potential of IoT. A novel consensus protocol named Proof of Understanding (PoU) with less energy consumption than PoW is proposed in Babelchain [59]. In this consensus protocol, a common message format is created approved by the sender and the receiver instead of solving a computationally hard puzzle.

1.4.1.5 Smart Contracts

A smart contract is a computer program that directly controls the transaction validation process of blockchain. Smart contracts in blockchain not only define the rules and penalties of an agreement but also enforce or self-execute contractual clauses. Cost reduction, precision, efficiency, and transparency are some of the benefits that encourage adopting it in many new applications. Therefore, smart contract can provide an efficient way to improve integrity and data security of the blockchain-based IoT systems.

Bitcoin uses a basic scripting language that validates a transaction based on certain rules including correct transaction format, valid signatures, and the fact that the transaction has not been previously spent [60,61]. On the other hand, Ethereum is the most prominent smart contract blockchain with a built-in Turing complete programming language that validates a transaction based on format, signatures, nonce, gas, and account balance of the sender's account [60]. IoT devices are heterogeneous and also vulnerable to cyber-attacks. In [62], the authors have identified that the fintech-oriented Bitcoin and general purpose Ethereum blockchain may not be suitable for IoT systems. Therefore, how the existing smart contracts can be applied in IoT systems is a big challenge.

1.4.1.6 Legal Issues

The blockchain technology is new in the industry and emerging with many possibilities day by day. While many are fascinated by the possibilities of blockchain, there are a lot of cautionary and contradictory statements from the experts. Specifically, the legality of the conversion of crypto currency to fiat money and vice versa has become a controversy. Though the European central bank has warned of legalization of virtual currency, it has also accepted it as a financial innovation. Therefore, prohibition of transactions with cryptocurrencies negatively impacts on blockchain uses. Also, the understanding of smart contract from a legal perspective is not straight-forward. Currently, the International Organization for Standardization (ISO) is working on understanding smart contracts from technical and legal perspectives [63]. In the context of IoT, a lot of legal issues are related to data privacy and protection law. Law regarding the IoT domain also differs from country to country. Therefore, it raises a big challenge for converging blockchain and IoT.

1.5 Chapter Summary

It is observed that the research on the convergence of blockchain and IoT is passing its early phase. Therefore, despite its promising benefits and bright future forecasts, it is facing initial development challenges. Being in the initial phase, a lot of research studies have been carried out in the field of smart homes, smart cities, and smart properties [63]. However, research on the convergence of blockchain and IoT is not limited within the field as mentioned earlier. For further development in this area, a roadmap to open research opportunities for the future is discussed below:

■ **Technical aspects**: In the beginning, blockchain was not solely developed for IoT platform. As time passes, the requirements of IoT-compatible blockchain are increasing. Though recently researchers are concentrating more on the technical side of blockchain and IoT, there are still issues regarding scalability, storage capacity, security, privacy, and consensus protocol that need huge research effort.

■ **Blockchain and cellular network for IoT**: The emerging IoT technology demands edge computing. A successful edge computing needs decentralized infrastructure that requires a distributed model of internet governance and decentralized trust. Nowadays, plenty of IoT edge relies on cellular networks. Research on decentralizing cellular networks is still in the nascent phase. Decentralizing the cellular network will not only benefit the security side but also provide flexible data packages and WiFi-sharing.

■ **Big data analytics for IoT**: IoT devices generate a huge volume of data in real time. This massive volume of data has huge business value. Applying big data analytics on IoT data can be very useful to retrieve hidden information and make decisions. But traditional big data analytics are not compatible with the IoT devices as they are resource constrained. Moreover, data analytics on anonymous blockchain data is difficult. Thereby, there is room for research in leveraging blockchain for big data analytics and IoT.

■ **IoT decentralization through IOTA or blockchain**: IOTA is a blockless cryptocurrency especially designed for IoT. It is a public and distributed ledger that employs the concept of "Tangle" [64]. Tangle is a new data structure based on Directed Acyclic Graph (DAG) that stores transactions instead of a conventional blockchain. IOTA addresses the issues of scalability, throughput, and latency problems that arise in blockchain technology. Therefore, it is believed that IOTA is the successor of blockchain technology. But some researchers from the MIT Media Lab were able to break into IOTA's customized hash function "Curl" which made its security questionable [65]. IOTA is specifically designed for IoT, but it is still under construction. Therefore, huge research efforts are needed in this field.

■ **Sharding in blockchain**: Sharding is a concept of database partitioning. It involves partitioning a large database into smaller ones to improve the performance of query response time. Currently, each node of a blockchain network needs to store the entire transaction information and processes all transactions. Therefore, presently the transaction confirmation time is very poor in blockchain. To reduce the transaction confirmation time, sharding in blockchain network is proposed [84]. The main idea is to break the blockchain network into multiple manageable shards to improve throughput and scalability issues. Like sharding in a database, sharding in a blockchain requires partitioning the blockchain network into multiple shards (segments). Each shard contains a unique transaction history. A subset of miner nodes is selected to process an individual shard. A careful selection of miner nodes is required to maintain security in the system. Zillica is the first public blockchain network that has implemented sharding [85]. Still, researches on implementing sharding are at the beginning level. A thorough study is required to discover its potential and future challenges.

References

1. K. Ashton, "That 'Internet of Things' Thing," *RFID Journal*, vol. 22, pp. 97–114, 2009.
2. R. Minerva, A. Biru, and D. Rotondi, "Towards a definition of the Internet of Things (IoT)," *IEEE IoT Initiative, Technical Report*, 2015, rev. 1.
3. J. Rivera and R. van der Meulen, "Forecast alert: Internet of Things endpoints and associated services, worldwide," 2016, Gartner (2016).
4. M. A. Khan and K. Salah, "IoT security: Review, blockchain solutions, and open challenges," *Future Generation Computer Systems*, vol. 82, pp. 395–411, 2018, ISSN 0167-739X.
5. Dr. Ovidiu Vermesan SINTEF, Norway, Dr. Peter FriessEU, Belgium, "Internet of Things–From Research and innovation to market deployment," river publishers' series in communications, 2014.
6. www.reloade.com/blog/2013/12/6characteristicswithin-internet-things-iot.php.
7. ITU, "Overview of the Internet of Things," *Ser. Y Glob. Inf. infrastructure, internet Protoc. Asp. next-generation networks - Fram. Funct. Archit. Model.*, p. 22, 2012.

8. A. Al-Fuqaha, M. Guizani, M. Mohammadi, M. Aledhari, and M. Ayyash, "Internet of Things: A survey on enabling technologies, protocols, and applications," *IEEE Communications Surveys & Tutorials*, vol. 17, no. 4, pp. 2347–2376, Fourthquarter 2015. doi:10.1109/COMST.2015.2444095.

9. R. Khan, S. U. Khan, R. Zaheer, and S. Khan, "Future internet: The Internet of Things architecture, possible applications and key challenges," *2012 10th International Conference on Frontiers of Information Technology*, Islamabad, 2012, pp. 257–260. doi:10.1109/FIT.2012.53.

10. T. Qiu, N. Chen, K. Li, M. Atiquzzaman, and W. Zhao, "How can heterogeneous Internet of Things build our future: A survey," *IEEE Communications Surveys & Tutorials*, vol. 20, no. 3, pp. 2011–2027, thirdquarter 2018. doi:10.1109/COMST.2018.2803740.

11. M. Khari, M. Kumar, S. Vij, P. Pandey, and Vaishali, "Internet of Things: Proposed security aspects for digitizing the world," *2016 3rd International Conference on Computing for Sustainable Global Development (INDIACom)*, New Delhi, 2016, pp. 2165–2170.

12. S. A. Kumar, T. Vealey, and H. Srivastava, "Security in Internet of Things: Challenges, solutions and future directions," *2016 49th Hawaii International Conference on System Sciences (HICSS)*, Koloa, HI, 2016, pp. 5772–5781. doi:10.1109/HICSS.2016.714.

13. Z. Yang, Y. Peng, Y. Yue, X. Wang, Y. Yang, and W. Liu, "Study and application on the architecture and key technologies for IOT," in *Proc. International Conference on Multimedia Technology (ICMT)*, Hangzhou, 2011, pp. 747–751.

14. M. Wu, T. J. Lu, F. Y. Ling, J. Sun, and H. Y. Du, "Research on the architecture of Internet of Things," in *Advanced Computer Theory and Engineering (ICACTE), 2010 3rd International Conference On*, Chengdu , 2010, pp. V5-484–V5-487.

15. E. Karafiloski, "Blockchain solutions for big data challenges A literature review," in *IEEE EUROCON 2017-17th International Conference on Smart Technologies*, Ohrid, 2017, no. July, pp. 6–8.

16. H. F. Atlam, A. Alenezi, M. O. Alassafi, and G. Wills, "Blockchain with Internet of Things: Benefits, challenges, and future directions," *International Journal of Intelligent Systems and Applications*, vol. 10, no. 6, pp. 40–48, 2018. doi:10.5815/ijisa.2018.06.05.

17. A. Banafa, "IoT and Blockchain Convergence: Benefits and Challenges," *IEEE IoT Newsletter*, 2017. [Online]. Available: http://iot.ieee.org/newsletter/january-2017/iot-and-blockchain-convergence-benefits-and-challenges.html.

18. M. S. Ali, M. Vecchio, M. Pincheira, K. Dolui, F. Antonelli, and M. H. Rehmani, "Applications of blockchains in the Internet of Things: A comprehensive survey," *IEEE Communications Surveys & Tutorials*. doi:10.1109/COMST.2018.2886932.

19. A. Reyna, C. Martín, J. Chen, E. Soler, and M. Díaz, "On blockchain and its integration with IoT. Challenges and opportunities," *Future Generation Computer Systems*, vol. 88, pp. 173–190, 2018, ISSN 0167-739X. doi:10.1016/j.future.2018.05.046.

20. A. Dorri, S. S. Kanhere, R. Jurdak, and P. Gauravaram, "Blockchain for IoT security and privacy: The case study of a smart home," *2017 IEEE International Conference on Pervasive Computing and Communications Workshops (PerCom Workshops)*, Kona, HI, 2017, pp. 618–623. doi:10.1109/PERCOMW.2017.7917634.

21. S. Sicari, A. Rizzardi, L. A. Grieco, and A. Coen-Porisini, "Security, privacy and trust in Internet of Things: The road ahead," *Computer Networks*, vol. 76, pp. 146–164, 2015.

22. R. Roman, J. Zhou, and J. Lopez, "On the features and challenges of security and privacy in distributed Internet of Things," *Computer Networks*, vol. 57, no. 10, pp. 2266–2279, 2013.

23. www.theguardian.com/technology/2016/oct/21/ddos-attack-dyn-internet-denial-service.

24. https://en.wikipedia.org/wiki/Information_security.

25. N. Komninos, E. Philippou, and A. Pitsillides, "Survey in smart grid and smart home security: Issues, challenges and countermeasures," *IEEE Communications Surveys & Tutorials*, vol. 16, no. 4, pp. 1933–1954, 2014.

26. S. Hilton, "Dyn analysis summary of Friday October 21 attack," http://dyn.com/blog/dyn-analysis-summary-of-fridayoctober-21-attack/, October 2016.

27. M. Antonakakis, T. April, M. Bailey, M. Bernhard, E. Bursztein, J. Cochran, Z. Durumeric, J. A. Halderman, L. Invernizzi, M. Kallitsis, D. Kumar, C. Lever, Z. Ma, J. Mason, D. Menscher, C. Seaman, N. Sullivan, K. Thomas, and Y. Zhou, "Understanding the mirai botnet," in *Proceedings of the 26th USENIX Security Symposium (Security)*, Vancouver, BC, 2017.

28. "3rd cyberattack 'has been resolved' after hours of major outages: Company," *NBC New York*, 21 Oct. 2016, bit.ly/2eYZO46.

29. N. Perlroth, "Hackers used new weapons to disrupt major websites across US," *New York Times*, 21 Oct. 2016, nyti.ms/2eqxHtG.

30. E. Blumenthal and E. Weise, "Hacked home devices caused massive internet outage," *USA Today*, 21 Oct. 2016, usat.ly/2eB5RZA.

31. M. U. Farooq, M. Waseem, A. Khairi, and S. Mazhar, "A critical analysis on the security concerns of Internet of Things (IoT)," *International Journal of Computer Applications*, vol. 111, no. 7, 2015.

32. H. Suo, J. Wan, C. Zou, and J. Liu, "Security in the Internet of Things: A review," in *International Conference on Computer Science and Electronics Engineering (ICCSEE)*, Hangzhou, 2012, vol. 3, pp. 648–651.

33. C. Kolias, G. Kambourakis, A. Stavrou, and J. Voas, "DDoS in the IoT: Mirai and other botnets," *Computer*, vol. 50, no. 7, pp. 80–84, 2017.

34. P. Tang and T. Venables, "'Smart' homes and telecare for independent living," *Journal of Telemedicine and Telecare*, vol. 6, pp. 8–14, 2000.

35. A. Dorri, S. S. Kanhere, and R. Jurdak, "Blockchain in Internet of Things: Challenges and solutions," *arXiv preprint arXiv:1608.05187*, 2016.

36. M. Vukolić, The quest for scalable blockchain fabric: Proof-of-work vs. BFT replication. In *Open Problems in Network Security*; Camenisch, J., Kesdŏgan, D., Eds; Springer: Cham, Germany, 2016; pp. 112–125. [CrossRef].

37. A. Dorri, S. S. Kanhere, R. Jurdak, and P. Gauravaram, LSB: A lightweight scalable blockchain for IoT security and privacy. *ArXiv, abs/1712.02969*, 2017.

38. J. D. Bruce, The Mini-Blockchain Scheme, 2014.

39. Mini-Blockchain Project, Cryptonite, 2014, http://cryptonite.info/and wiki: http://cryptonite.info/wiki/.

40. I. Eyal, A. E. Gencer, E. G. Sirer, and R. Van Renesse, "Bitcoin-NG: A scalable blockchain protocol," in *13th USENIX Symposium on Networked Systems Design and Implementation (NSDI 16)*, Santa Clara, CA, USA, 2016, pp. 45–59.

41. P. Ruckebusch, E. De Poorter, C. Fortuna, and I. Moerman, Gitar: Generic extension for Internet-of-Things architectures enabling dynamic updates of network and application modules, *Ad Hoc Networks*, vol. 36, pp. 127–151, 2016.

42. A. Taherkordi, F. Loiret, R. Rouvoy, and F. Eliassen, Optimizing sensor network reprogramming via in situ reconfigurable components, *ACM Transactions on Sensor Networks (TOSN)*, vol. 9, p. 14, 2013.

43. Filament, 2017. Accessed: 01 February 2018. Available online: https://filament:com/.

44. J. Barcelo, "User Privacy in the Public Bitcoin Blockchain," 2014. Last accessed: 12 December 2018 [Online]. Available: https://goo.gl/mN2y6V.

45. A. Biryukov, D. Khovratovich, and I. Pustogarov, "Deanonymisation of clients in bitcoin p2p network," in Proceedings of the ACM SIGSAC Conference on Computer and Communications Security, Scottsdale, Arizona, 2014, pp. 15–29.

46. M. Conoscenti, A. Vetrò, and J. C. D. Martin, 2016. "Blockchain for the internet of things: A systematic literature review," in *Proceedings of the IEEE/ACS 13th International Conference of Computer Systems and Applications (AICCSA)*, pp. 1–6. doi:10.1109/AICCSA.2016.7945805.

47. E. B. Sasson, A. Chiesa, C. Garman, M. Green, I. Miers, E. Tromer, and M. Virza, "Zerocash: Decentralized anonymous payments from bitcoin," in *Security and Privacy (SP), 2014 IEEE Symposium on*, San Jose, CA, USA, IEEE, 2014, pp. 459–474.

48. I. Miers, C. Garman, M. Green, and A. D. Rubin, "Zerocoin: Anonymous distributed e-cash from bitcoin," in *Security and Privacy (SP), 2013 IEEE Symposium on*, Berkeley, CA, USA, IEEE, 2013, pp. 397–411.

49. M. Schukat and P. Flood, "Zero-knowledge proofs in M2M communication,"' in *Proceedings of 25th IET Irish Signals & Systems Conference and China-Ireland International Conference on Information and Communications Technol*ogies, Limerick, Ireland, Jun. 2014, pp. 269–273.

50. Monero: Private Digital Currency, 2017. Last accessed 25 July 2018, https://getmonero.org/.

51. A. Kosba, A. Miller, E. Shi, Z. Wen, and C. Papamanthou, 2016. "The blockchain model of cryptography and privacy-preserving smart contracts," in *Proceedings of the IEEE Symposium on Security and Privacy (SP)*, pp. 839–858. doi:10.1109/SP.2016.55.

52. Hyperledger-Fabric Documentation, 2018. Last accessed 10 September 2018, https://media.readthedocs.org/pdf/hyperledger-fabric/latest/hyperledger-fabric.pdf.

53. Quorum-White Paper, 2016. Last accessed 5 July 2018, https://github.com/jpmorganchase/quorum-docs/blob/master/Quorum%20Whitepaper%20v0.1.pdf.

54. S. Nakamoto, 2008. Bitcoin: A Peer-to-Peer Electronic Cash System.

55. N. Szabo, "The idea of smart contracts," in *IEEE International Workshop on Electronic Contracting (WEC)*, San Diego, California, 2004.

56. R. Kastelein, Intel Jumps into Blockchain Technology Storm with 'Sawtooth Lake' Distributed Ledger, 2016. Last accessed 18 September 2018, www.theblockchain.com/2016/04/09/.

57. S. Popov, The Tangle, 2016, https://iota.org/IOTA_Whitepaper.pdf.

58. A. Baliga, 2017. Understanding Blockchain Consensus Models, White Paper, Persistent Systems Ltd.

59. Bitcoin Fog, 2016. Accessed: 2018-02-01. Available online: www:the-blockchain:com/2016/05/01/babelchain-machine-communication-proof-understanding-new-paper/.

60. V. Buterin, et al. A Next-Generation Smart Contract and Decentralized Application Platform, White Paper, 2014.

61. Bitcoin-Developer-Guide, Transactions, Developer Guide, 2018. Last accessed 13 September 2018, https://bitcoin.org/en/developer-guide#transactions.

62. I. Makhdoom, M. Abolhasan, and W. Ni, "Blockchain for iot: The challenges and a way forward," in *Proceedings of the 15th International Joint Conference on e-Business and Telecommunications - Volume 2: SECRYPT*, INSTICC, SciTePress, pp. 428–439, 2018. doi:10.5220/0006905605940605.

63. A. Panarello, N. Tapas, G. Merlino, F. Longo, and A. Puliafito, "Blockchain and IoT Integration: A Systematic Survey," *Sensors*, vol. 18(8), p. 2575, 2018.

64. What is Iota?, 2017. Last accessed 18 September 2018, https://iota.readme.io/v1.5.0/docs.

65. Iota Vulnerability Report: Cryptanalysis of the Curl Hash Function Enabling Practical Signature Forgery Attacks on the Iota Cryptocurrency, 2017. Last accessed 10 September 2018, https://github.com/mit-dci/tangled-curl/blob/master/vuln-iota.md.

66. Q. Jing, A. V. Vasilakos, J. Wan, J. Lu, and D. Qiu, "Security of the Internet of Things: Perspectives and Challenges," *Wireless Network*, vol. 20, no. 8, pp. 2481–2501.

67. N. Z. Aitzhan and D. Svetinovic, "Security and privacy in decentralized energy trading through multi-signatures, blockchain and anonymous messaging streams," *IEEE Transactions on Dependable and Secure Computing*, vol. 15, pp. 840–852, 2018.

68. M. S. Ali, K. Dolui, and F. Antonelli, "Iot data privacy via blockchains and ipfs," in *7th International Conference for the Internet of Things*, Linz, Australia, 2017.

69. O. Alphand, M. Amoretti, T. Claeys, S. Dall'Asta, A. Duda, G. Ferrari, F. Rousseau, B. Tourancheau, L. Veltri, and F. Zanichelli, "IotChain: A blockchain security architecture for the internet of things," in *Wireless Communications and Networking Conference (WCNC)*, 2018 IEEE. IEEE, Barcelona, Spain ,2018, pp. 1–6.

70. L. Axon and M. Goldsmith, Pb-pki: A Privacy-Aware Blockchain-Based PKI, 2016.

71. S.-C. Cha, J.-F. Chen, C. Su, and K.-H. Yeh, "A blockchain connected gateway for ble-based devices in the internet of things," *IEEE Access*, vol. 6, pp. 24639–24649, 2018.

72. G. Zyskind, O. Nathan, and A. Pentland, Enigma: Decentralized computation platform with guaranteed privacy, 2015. Last accessed 12 December 2018 [Online]. Available: https://enigma.co/enigma_full.pdf.

73. H. Shafagh, L. Burkhalter, A. Hithnawi, and S. Duquennoy, "Towards blockchain-based auditable storage and sharing of IoT data," in Proceedings of the Cloud Computing Security Workshop, Dallas, Texas, USA, 2017, pp. 45–50.

74. A. Ouaddah, A. Abou Elkalam, and A. Ait Ouahman, "Fairaccess: A new blockchain-based access control framework for the Internet of Things," *Security and Communication Networks*, vol. 9, no. 18, pp. 5943–5964, 2016.

75. K. Biswas and V. Muthukkumarasamy, "Securing smart cities using blockchain technology," in *IEEE 18th International Conference on High Performance Computing and Communications; IEEE 14th International Conference on Smart City; IEEE 2nd International Conference on Data Science and Systems*, Sydney, NSW, Australia , 2016, pp. 1392–1393.

76. B. Liu, X. L. Yu, S. Chen, X. Xu, and L. Zhu, "Blockchain based data integrity service framework for IoT data," in *IEEE International Conference on Web Services (ICWS)*, Honolulu, HI, 2017, pp. 468–475.

77. B. Lee and J.-H. Lee, "Blockchain-based secure firmware update for embedded devices in an Internet of Things environment," *The Journal of Supercomputing*, vol. 73, no. 3, pp. 1152–1167, 2017.

78. A. Boudguiga, N. Bouzerna, L. Granboulan, A. Olivereau, F. Quesnel, A. Roger, and R. Sirdey, "Towards better availability and accountability for IoT updates by means of a blockchain," in *IEEE European Symposium on Security and Privacy Workshops (EuroS&PW)*, Paris, 2017, pp. 50–58.
79. M. Steger, A. Dorri, S. S. Kanhere, K. Römer, R. Jurdak, and M. Karner, "Secure wireless automotive software updates using blockchains: A proof of concept," *Advanced Microsystems for Automotive Applications*, vol. 2018, pp. 137–149, 2017.
80. Z. Yang, K. Zheng, K. Yang, and V. C. Leung, "A blockchain-based reputation system for data credibility assessment in vehicular networks,"*2017 IEEE 28th Annual International Symposium on Personal, Indoor, and Mobile Radio Communications (PIMRC)*, Montreal, QC, 2017, pp. 1–5.
81. A. Bahga and V. K. Madisetti, "Blockchain platform for industrial internet of things," *Journal of Software Engineering and Applications*, vol. 9, no. 10, p. 533, 2016.
82. R. B. Chakraborty, M. Pandey, and S. S. Rautaray, "Managing computation load on a blockchain-based multi-layered Internet-of-Things network," *Procedia Computer Science*, vol. 132, pp. 469–476, 2018.
83. N. Kshetri, "Can blockchain strengthen the Internet of Things?" *IEEE Computer Society*, August, vol. 19, no. 4, pp. 68–72, 2017.
84. R. James. On Sharding Blockchains, 2018. Last accessed 15 September 2018, https://github.com/ethereum/wiki/wiki/Sharding-FAQ.
85. www.iot-now.com/2018/04/17/80758-zilliqa-releases-first-sharding-blockchain/.

Chapter 2

IoT Blockchain Integration: A Security Perspective

Kazım Rıfat Özyılmaz and Arda Yurdakul

Bogazici University

Contents

2.1 Introduction

Internet of Things (IoT), which is the term that is used to define the whole set of connected devices, is continuously expanding at an increased rate. In 2016, it was forecasted that total number of IoT devices will reach 30 billion by 2020 and 75 billion by the end of 2025 [1]. However, research shows that there were only 17.8 billion connected devices in 2018, of which only 7 billion are IoT devices. Nevertheless, current expectation is 9.9 billion IoT devices for 2020 and 21.5 billion IoT devices in 2025 [2]. Although the aspect of being connected is the main property of IoT devices, a much concise definition is given by Madakam et al. [3]:

> An open and comprehensive network of intelligent objects that have the capacity to auto-organize, share information, data and resources, reacting and acting in face of situations and changes in the environment.

This definition emphasizes other important aspects of IoT infrastructure such as the capability to interact and adapt. The security of these billions of highly adaptive devices is another problem. For example, recent attacks to DNS infrastructure by the Mirai Botnet demonstrated how easily IoT infrastructure could be compromised [4].

Statistics show that most deployments of the IoT projects in 2018 have focused on "Smart City," "Smart Industry," and "Smart Building." Especially, traffic, utilities, and lighting are the main topics that the IoT solutions are targeting [5]. As expected, the first step of improving conditions in a densely populated area should be monitoring the utilization of the public resources. Wide-scale deployment of IoT devices with their decreasing costs and increasing coverage will present a solid option for municipalities and governments. With such increased measurement capabilities, local authorities will have a wide range of improvement and optimization opportunities starting from sanitation services, parks and recreation services to public security or rearranging of public places based on utilization. In addition, services on the ground will not be the only ones to benefit from the adoption of IoT devices. With smart grid, all the infrastructure services like energy and water will be continuously monitored. This way, resource efficiency will be increased and reaction times to any fault will be significantly decreased or much better anticipated before it happens.

Today, the world is revolving around data. We expect smarter devices and services that anticipate our needs and requirements instantly. Thanks to IoT devices and the huge amount of continuously collected data, the industry is working on creating predictive services for all kinds of instances, although it is mostly offline. These capabilities bring up a new set of questions: where and when is this information collected? How is it stored? How is that particular piece of data processed? Finally, how is that precious insight deduced? As seen, data is a multifaceted concept: in order to create any insight from a piece of data, it should go through a manufacturing-like process. The fundamental concepts of this pipeline can be given as follows: data collection, data transfer, data storage, and data processing. Obviously, such a pipeline instantly spawns a lot of important questions: who owns that data, does it comply with the regulatory privacy requirements like General Data Protection Regulation (GDPR) (Regulation (EU) 2016/679) [6], and do the monetization strategies of the service conflict with any of the above? To sum up, IoT solutions are focusing on creating and understanding data to generate actionable insights. Any IoT solution that deals with data will inherit all of the benefits and liabilities attached to it.

In order to understand and improve current IoT solutions, existing technology stack should be examined. Modern IoT systems consist of a perception layer that collects data from the environment, a network layer that deals with networking components like gateways or cloud backends, and application layer that implements the user-facing product [7]. With the influence of semantic web technologies and big data, new data-centric layers such as learning layer and action layer are also proposed [8]. Drastic technological innovations may be observed for both perception and network layers. For example, new low-power sensors are getting smaller and cheaper day-by-day, and new communication technologies like low-power wide-area networks (LPWANs) and next-generation cellular mobile communication technologies (5G) are rapidly emerging. Similarly, with new computing trends, like edge computing, computation is pushed more into the network layer.

Major cloud providers like Amazon Web Services are providing backend solutions to increase adoption and ease on-boarding complexity [9]. However, are the device manufacturers and system integrators ready for the exponentially increasing device deployment in terms of backend and rollout capabilities? Cloud providers solved the scalability issue at least for their own data center domain, but every IoT manufacturer or system integrator should solve the scalability and security problem alone. This means that every IoT company should define and implement a continuous process to manage cloud resources dynamically while maintaining secure data flow and storage. Basically, this task is neither straightforward nor easy as dynamic cloud resource configuration and heterogeneous IoT device profile management still pose a significant challenge. In addition, interoperability and standardization in terms of communication protocols and data syntax are highly important [7,10].

Blockchain infrastructure may provide a much bigger opportunity than only powering a digital currency: creating decentralized trustless systems. The impact of blockchain technology may be better understood by quoting how IBM is envisioning the future of IoT [11]:

> In our vision of a decentralized IoT, the blockchain is the framework facilitating transaction processing and coordination among interacting devices. Each manages its own roles and behavior, resulting in an 'Internet of Decentralized, Autonomous Things' and thus the democratization of the digital world (…)
>
> In this democracy of hundreds of billions, users bind with devices using secure identification and authentication (…) Devices, on the other hand, are empowered to autonomously execute digital contracts such as agreements, payments and barters with peer devices by searching for their own software updates, verifying trustworthiness with peers, and paying for and exchanging resources and services.

Just as IBM has envisioned, various products and services that integrate IoT and blockchain are in motion. Examples may include a blockchain platform for on-demand manufacturing [12], an API-based blockchain-as-a-service (IOTA) platform [13], a blockchain-based IoT security solution [14], and even a smart lock product [15]. On a grander scale, blockchain-based grid solutions like Brooklyn Microgrid [16] provided by LO3 Energy pave the way to decentralized energy metering solutions. Besides industry, academia is also focusing on various IoT and blockchain integration use-cases. Examples include managing IoT devices using a blockchain platform [17], study of IoT privacy and security in a smart home environment [18], or increasing IoT data privacy via blockchain and Interplanetary File System (IPFS) [19].

In the end, blockchain technology may help propelling IoT to the next level with its open infrastructure, where data integrity, method of transfer, and ownership are protected by cryptographic primitives. Basically, blockchain will provide a general data storage and management infrastructure where data interfaces are clearly defined, documented, and implemented. Such a standardization will strengthen the security of the system and minimize the attack surface. On top of the security benefits of standardization, blockchain has already been proposed for storing safety critical information like firmware signatures, binary checksums, and malicious software indices for IoT devices [20,21]. This way, IoT devices may use a public, immutable decentralized database to verify their applications and be safe against known malicious actors.

2.2 Chapter Roadmap

This chapter is divided into eight sections where each section discusses a certain aspect of the IoT–blockchain integration and its security implications.

In Section 2.3, we will present emerging directions and paradigm shifts in IoT technologies which are mainly in the information processing and networking domains.

Recent trends like fog computing, LPWANs, 5G networks, and the use of artificial intelligence (AI) and machine learning (ML) in IoT will be the focus of the chapter.

Security, privacy, and scalability are the main issues that must be considered for mass deployment and adoption of IoT technology. We believe that a blockchain-based IoT infrastructure will solve most of these issues. Hence, Sections 2.5 and 2.6 are devoted to explain how blockchain can solve security and privacy problems of IoT. First, we present "Blockchain Concepts" (Section 2.4) to set forth the fundamental concepts in blockchain technologies such as digital signatures, public-key cryptography, directed acyclic graphs (DAGs), and consensus functions. On top of that, security and privacy technologies like zero-knowledge proofs, confidential transactions, and ring signatures have been discussed and a brief overview of the current distributed filesystems and application development platforms is given. The security implications of using blockchain in IoT are detailed in Section 2.5, as we show how blockchain will be a promising infrastructure for IoT systems. This section focuses on security mechanisms and countermeasure techniques provided by blockchain technologies at device, network, and application levels.

Blockchain is famous for being used as the infrastructure of cryptocurrencies, like Bitcoin and Ethereum, which rely on powerful computing machines. However, in the scope of IoT, various types of devices will be deployed in the field. In Section 2.6, we evaluate the integration methods of IoT devices with varying power and computational capabilities to a blockchain-based infrastructure. In this section, we classify the devices in an IoT deployment according to their power consumption to show that they may need to be enhanced with blockchain-specific features ranging from a full blockchain miner node to a plain data source interfacing the blockchain-based IoT infrastructure. We also present three different IoT–blockchain use-cases in Section 2.7 to demonstrate how a blockchain-based IoT infrastructure can be combined with the emerging trends in IoT. In addition, the standardization of the programming interfaces to store and access data, the utilization of a decentralized data storage, and their combined applications are the main takeaways from this section.

Finally, in Section 2.8, we provide a summary of the chapter with the key insights that can be gained from this chapter.

2.3 Emerging Trends in IoT

IoT technologies as a whole are rapidly evolving. As expected, IoT deployment architectures and corresponding technology stacks are affected by significant changes in computing paradigms and emerging communication standards. In addition, recent advances like "deep convolutional neural networks" and production of application-specific integrated circuits (ASICs) for ML make it possible to integrate intelligence to IoT applications. In this section, we will list the major paradigm shifts in IoT technologies and go over their impact on security and privacy.

There are two major trends that may affect the organization of the IoT networks: the first one is the edge computing paradigm [22], and the second one is the emerging communication technologies like LPWANs and 5G mobile networks [23]. In the crossroads of both paradigms, it becomes clear that mass IoT deployments in the future will consist of heterogeneous devices. With the range and low-power capabilities of LPWAN technologies, a star topology network and a gateway-centric deployment will be adopted. For large-scale IoT deployments where coverage and battery life are the most important parameters, transition from peer-to-peer sensor networks to LPWANs should be expected [24].

As discussed in the Introduction, the goal of IoT is not solely focused on measurement and monitoring. Creating a usable insight from data is the ultimate goal. In today's worlds, such a system should definitely benefit from the achievements in the ML field. Vast research on both deep learning and convolutional neural networks is present, and even industry is sharing algorithms and designing dedicated hardware to accelerate such technologies. The quality of an application of such technologies is correlated with the amount of data that is consumed both at training and correction phases, so IoT is one of the great data resources that will help in the creation of high-quality predictive and preventive applications.

2.3.1 Edge and Fog Computing

Since its inception in 2005, cloud computing changed the way that modern services are developed, deployed, and interacted with their consumers. However, a cloud backend that stores and processes data sent from the end devices and serves connected users has started to become suboptimal due to the increase in the number of IoT devices. Shi et al. [22] state that there are three major reasons that trigger the requirements for a new kind of computational workload distribution. The first reason for that is it is no longer feasible to push all data to cloud due to the bandwidth bottlenecks. The amount of connected devices and the quantity of raw data produced are increasing much faster than the bandwidth capabilities of the infrastructure. The second one is, in a connected environment (home or industrial), data is mostly consumed at the local network where it is acquired. By sending the data to the cloud for processing and getting results back, the latency and the response time of systems are actively increasing. Lastly, the direction of data is not unidirectional anymore. IoT devices today are not only producers of data but also its consumers.

2.3.2 Low-Power Wide Area Networks

Modern IoT applications have a variety of requirements such as low data rate, long range, low power, and being cost-effective. Low-power technologies like Bluetooth or Zigbee are unable to provide long-range data transmission capabilities. On the other hand, established long-range communication technologies like cellular

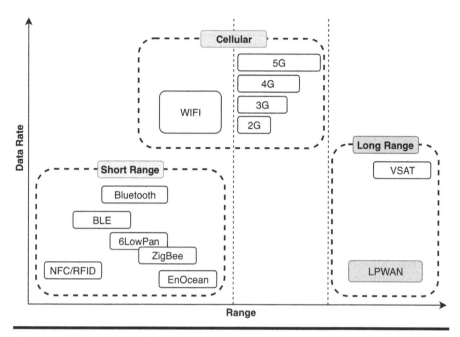

Figure 2.1   LPWAN positioning [23].

systems (e.g., 2G, 3G, and 4G) are consuming too much power to be used in an IoT deployment. A new communication technology, namely, LPWAN, have emerged because of these requirements [23]. As seen in Figure 2.1, LPWAN technologies have low data rate compared to cellular technologies; however, the communication range is significantly higher than its counterparts.

LPWANs are categorized as licensed and unlicensed based on their operating frequencies. Licensed LPWANs provide good quality of service (QoS) and operate well in populated areas. Examples for licensed LPWANs may include eMTC (LTE Cat.M), EC-GSM, and NB-IoT [25]. In contrast, unlicensed LPWANs have no guaranteed latency [26] and lower QoS but they provide better coverage, lower power requirements, and lower device costs. Comparison of these technologies is given in Table 2.1 [27]. Among unlicensed technologies, LoRa has better data rate compared to Sigfox with only a small decrease in range.

In terms of technology differences, LoRa, Sigfox, and NB-IoT technologies may be evaluated by the following categories: QoS, battery life, scalability, network coverage, deployment model, and cost. As said, NB-IoT employs a licensed spectrum and therefore provides a guaranteed QoS. On the other hand, LoRa and Sigfox are the best options for better battery life due to synchronous communication and QoS handling of NB-IoT. NB-IoT both offers the advantage of maximum payload length of 1,600 bytes compared to LoRa's 243 bytes and Sigfox's 12 bytes and better scalability (100K end devices per cell) compared to

Table 2.1 LPWAN Connectivity Overview [28]

	LoRa	Sigfox	NB-IoT (Rel.13)	eMTC (Rel.13)	EC-GSM (Rel.13)
Range	<11 km	<13 km	<15 km	<11 km	<15 km
Max coupling loss	157 dB	160 dB	164 dB	156 dB	164 db
Spectrum	Unlicensed <1 GHz	Unlicensed 900 MHz	Licensed LTE	Licensed LTE	Licensed GSM
Bandwidth	< 500 kHz	100 Hz	180 kHz 200 kHz Carrier	1.08 MHz 1.4 MHz Carrier	200 kHz
Data rate	<50 kbps	<100 bps	<170 kbps (DL) <250 kbps (UL)	<1 Mbps	<140 kbps

LoRa and Sigfox's 50K. As given in Table 2.1, coverage of Sigfox and LoRa is close but it is much better than NB-IoT. In terms of deployment, LoRa gives the option to deploy a local network; therefore, it is more powerful and flexible than Sigfox and NB-IoT. Lastly, NB-IoT is by far the most costly option for the network operator due to the spectrum licenses (>500M Euro/MHz) and per base station cost. NB-IoT base stations cost around 15,000 Euro where it is around 4,000 Euro for Sigfox and around 1,000 Euro for LoRa. Both LoRa and Sigfox have very affordable end devices (<5 Euro) compared to NB-IoT (20 Euro) end devices (Figure 2.2) [23].

As discussed, LPWANs are gateway-centric communication structures, where gateways are more resourceful in terms of memory and computation capabilities. This property makes LPWAN-based IoT deployments very suitable to blockchain integration (Section 2.8).

2.3.3 5G Networks

The fifth-generation cellular mobile communication technology (5G) will provide architectural improvements for IoT systems. Li et al. [29] provide a list of key enabling technologies that effect IoT systems in particular. 5G networks have a modern architecture that relies on software defined networks (SDN). Therefore, they offer Wireless Network Function Virtualization (WNFV), Heterogeneous Networks (HetNet), and Direct Device to Device (D2D) technologies, all of which are promising a better communication infrastructure for IoT devices. By using

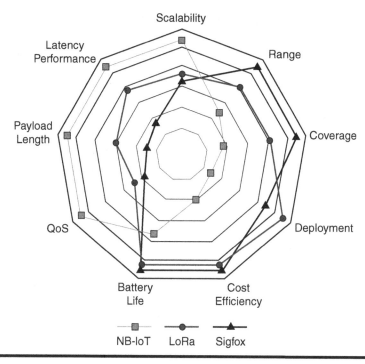

Figure 2.2   Respective advantages of Sigfox, LoRa, and NB-IoT [23].

HetNet networking paradigm, 5G network will be capable of implementing on-demand information transmission rates. In addition, with the help of D2D technology, a method of data transmission for short range is proposed. This way, it would be possible to get better QoS even while using low-power devices.

Due to the properties like HetNet and increased bandwidth capabilities, 5G IoT solutions may need additional care in terms of security especially for certain types of attacks like authentication, access control, communication channel protection, confidentiality, integrity, and availability. Tiburski et al. [30] evaluated various IoT 5G middleware software packages (VIRTUS, ONEM2M, SOCRADES, COSMOS, SIRENA, and HYDRA) and inspected whether preventive security mechanisms for the listed attacks are implemented. Based on that research, no 5G IoT middleware is ready in terms of security, except the ONEM2M software.

2.3.4 Machine Learning and Artificial Intelligence

According to McKinsey, ML and AI are the two leading research topics that became especially popular in 2018 [31]. By using mathematical models that are composed of multiple processing layers (deep learning), the performance of pattern recognition tasks is significantly improved and these techniques are widely adopted in various applications today [32]. Ensuring security in computer networks is one of the application areas of deep learning technology [33].

Security comes up as a two-faceted concept in ML. First one, as expected, is the security of the actual ML system. This will cover attacks to the training system, its training data, its classification mechanism, and corresponding actions. Barreno et al. [34] classify these types of attacks in the three axes: influence, security violation, and specificity. Attackers may influence learning by either manipulating training data or exploiting misclassifications. In addition, security of the ML system may be violated by compromising assets via using false negatives. Similarly, Denial of Service (DoS) attacks may be conducted by using false positives. Attacks against these systems may also focus on a single user or a broad range of instances. In addition to DoS attacks, targeted attacks may occur in order to break up anonymization of a certain user or a selected group of users.

Ensuring cyber security is the second aspect that comes with ML. ML algorithms combined with data mining techniques open up a lot of interesting possibilities for cyber security applications, especially intrusion detection. Buczak et al. [35] provide information on data set properties, possible ML algorithms, and their evaluation results.

In this paper, ML will be evaluated as a complementary technology, which may be coupled with blockchain to enhance the security of IoT platforms (Section 2.8).

2.4 Blockchain Concepts

Today, all kinds of different blockchain protocols are proposed, and the focus of these projects varies greatly. Nonetheless, cryptography is at the center of all these new protocol efforts, and in this section, the fundamental concepts used in blockchain systems will be presented.

2.4.1 Digital Signatures and Public-Key Cryptography

Digital signatures and public-key cryptography are the foundation of the blockchain technology. Digital signatures are invented to communicate information on unsecured channels by creating a signature using a key pair (one public and one private). There are three major components of a digital signature algorithm: key generation, signing, and verification.

The first notion of digital signature algorithms is attributed to Whitfield Diffie and Martin Hellman [36]. The underlying mechanism is a one-way trapdoor function which is easy to compute in one direction, but extremely difficult to calculate in the reverse direction. Soon after that, Ronald Rivest, Adi Shamir, and Len Adleman invented the RSA algorithm [37]. Other notable mentions on digital signatures are Lamport signatures [38] with non-reusable keys and Merkle trees [39], which are extensively used in blockchain systems for transaction verification.

On top of these cryptographic functions and primitives, modern blockchain systems use elliptic curve cryptography (ECC) [40,41] to sign and verify transactions, because the keys that can be used in ECC may be much more smaller relative to RSA keys. For example, ECC may use 160 bits for the security level provided by RSA with 1,024 bits. Therefore, the amount of signature data that should be stored in the blockchain is significantly smaller.

2.4.2 Blockchain Fundamentals

Blockchain is a distributed database where a replica of the whole database is stored in every participating *node*. By definition, it is a chain of *blocks* that are cryptographically linked and timestamped collections of transactions. Continuous verification of blocks and transactions is an essential component of blockchain systems to eliminate malicious attacker's attempts of forging transactions. Public-key cryptography and hash algorithms are at the heart of blockchain technology [42].

As seen in Figure 2.3, every block has a header that contains a timestamp value, the hash value of the previous block's header, and the hash value of all the transactions in a Merkle tree form [43]. By using these two hash values, integrity of all the carried transactions and the block header can be independently verified.

In blockchain-based systems, the consensus mechanism will decide which transactions will be included to the ledger, i.e., blockchain. *Nakamoto Consensus* proposed a computational way to do that which is called *Proof-of-Work* (PoW), where any peer in the system may broadcast a valid block if the computational requirement is satisfied [42]. Block creation will work as follows: first, *miners* will build a block with available transactions and calculate a hash value for the

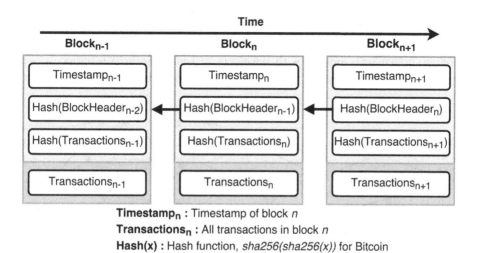

Figure 2.3   Blockchain structure.

block header. This calculation is a *one-way function* which is easy to verify but extremely hard to compute the other way around. Second, the resulting hash value is compared to the *difficulty* value of the system at the time, and if the hash is smaller than the difficulty parameter, the proposed block will be deemed valid. Lastly, this valid block will be sent to the network and all the other peers independently verify and accept it.

2.4.3 Directed Acyclic Graph (DAG)

DAG is a special kind of graph where all edges are directed and no cycles exist (Figure 2.4). Therefore, a single node may be pointed by multiple previous nodes, creating an interconnected state graph ordered by time. Based on these properties of DAG, a new concept called Tangle [44] is introduced as an alternative to blockchain systems. Instead of creating a chain of cryptographically verifiable blocks, transactions are "entangled" to each other and a web of connections are created, where every transaction refers to a group of previously validated transactions.

IOTA [13], which is an open source distributed ledger project targeting IoT devices, adopted Tangle technology in order to increase scalability and reduce latency. Despite its advantages, increased ledger size is a significant problem for DAG-based distributed ledgers. Current implementations (like IOTA) introduce a mechanism called "snapshots" which purge transaction history for regular nodes while keeping the complete history only at special nodes called "permanodes."

2.4.4 Consensus Algorithms

Consensus algorithm (or function) is the core mechanism that peers in a distributed network use to agree on the value of a piece of data. Consensus functions may be classified into two major categories: traditional consensus and Nakamoto consensus that is introduced with Bitcoin [42]. Today, blockchain protocols use various consensus algorithms, which in turn defines how decentralized or trustless

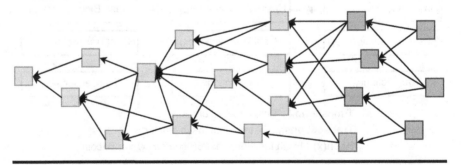

Figure 2.4   Directed acyclic graph.

their system will be. In this chapter, major consensus algorithms in the blockchain space will be presented.

Proof-of-Work (PoW) or Nakamoto consensus proposes a computationally verifiable way of coming to a consensus in a distributed system. At its core, PoW uses a mechanism called one-way function. In terms of computational resources, these functions are very cheap to calculate in one direction, but extremely expensive in reverse. PoW forces peers to prove that a certain amount of computational resources are utilized (by satisfying a difficulty parameter) when proposing a new block. The difficulty value is adjusted based on the block creation times [42]. By design, it takes 10 minutes and 17 seconds to create a valid block in Bitcoin and Ethereum, respectively.

Proof-of-Stake (PoS) is a consensus protocol that aims to create voting rights for the next block based on the distribution of the asset class (token, coin, etc.) in the system. In PoS, every participant in the system may have the chance to create the next block (or at least help to create the next block) according to its financial commitment in the consensus process. Peercoin [45] introduced the concept of PoS, and it was highly popular due to the fact that it does not consume energy like PoW schemes. Ethereum is also planning to shift to PoS consensus mechanism with the "Casper" upgrade [46].

Delegated Proof-of-Stake (dPoS) is a version of PoS protocol where participants vote for a limited number of special users, namely, witnesses, who are responsible for creating new blocks. In addition, participants select another set of users called delegates who are focusing on governance and performance aspects of the system [47]. dPoS is criticized because of the centralization aspect in terms of mining, and it is currently used in projects like Lisk [48], EOS [49], Steem [50], and BitShares [47].

Practical Byzantine Fault Tolerance (PBFT) is a traditional consensus algorithm that is based on state-machine replication for distributed systems [51]. As expected, PBFT is designed for reaching consensus with unreliable actors, and it may tolerate up to one-third of the total nodes being malicious [52]. In practice, all transactions coming from the peers are signed, and once the number of signatories reaches a certain threshold, transaction is deemed valid. One interesting property of PBFT is satisfying finality, which means it does not allow temporary forks like the other consensus algorithms [53]. PBFT is selected as Hyperledger Fabric's default consensus algorithm [54].

Proof-of-Authority (PoA) is a simple consensus protocol designed to be used in private blockchains. PoA relies on a set of block creators selected at genesis block. PoA allows to set a percentage threshold for block validity; i.e., if more than a certain percentage of the creators are signed, the block is valid. It is also possible to add additional verifiers via a voting mechanism. PoA is a system governed by trusted parties so the security of the whole network is tied to the security of each trusted party individually. Currently, PoA is used in Ethereum's biggest test network Rinkeby [55].

Proof-of-Elapsed-Time (PoET) is not a major consensus algorithm; however, it is worth mentioning because it provides an improvement of efficiency by using a *trusted execution environment (TEE)* [56]. PoET consensus elects peers to execute requests, and the peer with the smallest response time is the winner. However, peers have to wait for an amount of time, which is determined by a exponentially distributed random variable. The whole process relies on TEE, and it is verifiable that no cheating has occurred. Currently, PoET is used in Hyperledger Sawtooth, and it uses Intel SGX Enclave as the TEE [56]. Unfortunately, there has been a significant research on various aspects of Intel SGX Enclave like cache attacks [57] and out-of-order execution-related security problems [58].

Avalanche is a new type of consensus algorithm which introduces a new family of Byzantine fault tolerance protocols. Avalanche steers all nodes in the system to a consensus via voting and queries, but it does not need a leader to do so. Avalanche protocol starts with *Slush* stage where nodes in the network query each other, and if the majority is picking a state (it is expressed as a *color* in the paper), the nodes slowly change their state based on network responses. *Snowflake* is the part, where historic information of these choices is added. In *Snowball*, a level of confidence is assigned to those history of choices. Finally, *Avalanche* generalizes Snowball and creates a DAG of all known transactions [59]. In terms of security, Avalanche consensus is still under development and there is no large-scale experimental data.

2.4.5 Zero-Knowledge Proofs

Zero-knowledge proof is the method of proving the ownership of a piece of data (the prover) to another party (the verifier), without conveying any other information other than the correctness of the proposition in question [60]. A zero-knowledge proof must satisfy the following requirements: completeness, soundness, and zero-knowledge. Basically, an honest verifier will be convinced by an honest provider if the proposition in question is true (completeness). However, it is also not possible to cheat the verifier if the proposition is false (soundness). Finally, it is also not possible to extract any other information except that the proposition is true (zero-knowledge).

Zero-knowledge proofs are becoming widely used in blockchain domain for ensuring both privacy and security. By using zero-knowledge proofs, it is possible to create non-traceable transactions and design a new kind of secure authentication systems. A list of projects that leverage this technology are as follows:

- Zcash implemented a digital currency with strong privacy guarantees, leveraging the recent advances in *zero-knowledge Succinct Non-interactive ARguments of Knowledge (zk-SNARKs)* [61].
- *Bulletproofs* is a new non-interactive zero-knowledge proof protocol without a trusted setup that has very short proofs [62].

- *Mimblewimble* is a blockchain protocol focused on privacy, scalability, and fungibility in digital transactions [63].
- *Anonymous Zero-knowledge Transactions with Efficient Communication (AZTEC)* protocol is aiming to enable private transactions on Ethereum [64].

2.4.6 Confidential Transactions and Ring Signatures

Confidential Transactions are a special kind of transaction that aims not to expose any information on either the exchanged value (may be a measurement data), identities of transacting parties, or both. The idea behind confidential transactions was proposed by Adam Beck for Bitcoin, and it was formally defined by Greg Maxwell [65].

Monero, a privacy-oriented cryptocurrency, extended confidential transactions by combining it with ring signatures [66]. Basically, any member in a particular group may create a ring signature for a transaction, and it is computationally infeasible to find out that particular member. By combining both of these technologies, Monero introduced *RingCT* [67]. In RingCT, it is not possible to trace the source, destination, and sum in the transaction. Similar to zero-knowledge proofs, these technologies are used to enhance privacy and security.

2.4.7 Distributed Filesystems

Distributed filesystems are peer-to-peer, fault-tolerant, zero-downtime, DDOS, and censorship-resistant storage services. Examples for such torrent-like services may include IPFS [68], Swarm [69], and Storj [70]. By coupling decentralized storage with blockchain, it is possible to use off-chain, decentralized storage and verify the integrity of the stored data using on-chain transaction history.

2.4.8 Distributed Applications (DApps)

In blockchain domain, *DApps* are applications that are stored on blockchain and run on multiple systems simultaneously. The most widely known platform for distributed apps is Ethereum. On Ethereum, it is possible to develop *smart contracts* (with a custom programming language) that will be executed via *Ethereum Virtual Machine* (*EVM*). DApps on blockchain provide two significant advantages:

- Smart contracts will be deployed on all nodes so it is zero-downtime by design.
- It is not possible to modify the deployed smart contract on blockchain; therefore, it is tamper-proof.

The most significant downside of DApps is that they are vulnerable to EVM-related attacks. Once an EVM vulnerability is disclosed, any smart contract that has the flaw may be exploited [71].

2.5 IoT–Blockchain Promises

Integrating blockchain systems with IoT will bring up a whole new set of tools to address security and privacy issues in IoT. In the first part of this chapter, the taxonomy of security issues in IoT will be given and the built-in security mechanisms of blockchain will be evaluated. Such an evaluation will help in mapping the security and privacy issues to solutions coming from blockchain's fundamental properties. In the second part, various IoT–blockchain use-cases will be presented and an in-depth analysis will be given to show how blockchain makes IoT systems more secure, autonomous, and decentralized.

2.5.1 Blockchain Security Mechanisms

Khan et al. [72] present a taxonomy for IoT security issues and divide them into three categories: low-level, intermediate-level, and high-level security issues. In addition, the security threats and the proposed solutions in these categories are further grouped using the affected network layers like physical, transport, and application. Based on these definitions, Table 2.2 provides ways to combat device-, network-, and application-level security issues using blockchain technology.

Based on the IoT security model, blockchain infrastructure will provide solutions to security issues in three levels: device level, network level, and application level. At device level, any attempt to modify the software on the device can be detected by cross-referencing software signatures with the ones stored in blockchain.

At network level, blockchain systems present two options to overcome attacks: the first one is using countermeasures for not getting identified as a target, and the second one is using authentication and blacklisting mechanisms to exclude malicious peers. For example, Dandelion [73] is a peer-to-peer communication protocol for Bitcoin that is designed not to leak any identifying data. On the other hand, attacks like Sybil, DoS, or IP spoofing can easily be detected if a blockchain-based authentication service is used [54]. Furthermore, it won't be possible to execute a successful replay attack because all the blockchain platforms are resistant to the double-spend problem.

Lastly, at application level, blockchain systems will provide a lot of powerful tools and services, especially to protect the peer's privacy. Similar to device-level countermeasures, authenticity of any application may be verified by cross-referencing its signature with the data stored in blockchain. In addition, it is possible to use crpytographic techniques like zero-knowledge proofs, confidential transactions, or ring signatures to conceal sensitive data.

2.5.2 Security and Privacy Enhancing Use-Cases

Blockchain as the trust mechanism: Blockchain systems use cryptographic primitives and consensus functions to provide its users the capability to interact with each other without the need of a trusted third party. In practice, this property of

Table 2.2 IoT–Blockchain Security Countermeasures

IoT Levels	Security Issues	Blockchain Countermeasures	Sample Implementations
Device level	Insecure physical interface	Software signature verification	All immutable ledgers
	Insecure initialization and configuration	Software signature verification	All immutable ledgers
Network level	Sybil attacks	Peer authentication, P2P countermeasures	Hyperledger [54]
	DoS attacks	Peer authentication, P2P countermeasures	Hyperledger [54]
	IP spoofing	Peer authentication, anonymization	Dandelion protocol
	Replay attacks	Replay protection	All immutable ledgers
Application level	Insecure (password) interfaces	Zero-knowledge proofs	Zcash [61]
	Insecure software/firmware	Software signature verification	All immutable ledgers
	Middleware security	Software signature verification	All immutable ledgers
	Privacy violation	Zero-knowledge proofs, confidential transactions	Zcash [61], Monero [67]

blockchain should pave the way for the rise of the completely autonomous IoT devices and networks.

Blockchain as the common repository: As discussed in the edge computing section, cloud bandwidth and its communication latency will not be able to sustain the exponentially rising IoT device connections. However, a global, decentralized, and secure data repository may standardize the way data is pushed in and help in minimizing the attack surface.

Blockchain as the AI/ML integration layer: In order to produce good actionable insights, AI and ML models need lots of data for training. However, data is harvested by big companies and kept in locked data silos today. If a blockchain-based, common data repository ever exists, it is only logical that it becomes the integration layer for AI and ML systems. In terms of security and privacy, it may finally be possible to build transparent and auditable systems where the whole process from data acquisition to actionable insight creation is followed.

Blockchain as the application environment: Despite their current limitations, having a blockchain-based decentralized application infrastructure minimizes the attack vector in a significant manner. Due to the fact that the application itself is stored on blockchain, it is not possible to alter the application.

2.6 IoT–Blockchain Integration

IoT end devices and gateways vary greatly in terms of power requirements and computational capabilities. In terms of power requirements, IoT end devices may be either battery-powered or connected to a fixed power line. On the other hand, typical IoT gateway (especially in a LPWAN scenario) is almost always connected to a fixed power line. As a result, integration with a blockchain infrastructure requires multiple roles that can be used for a spectrum of IoT devices. Figure 2.5 presents an integration approach.

Gateway as a full blockchain node: In this scenario, IoT gateways should have the bandwidth, storage, and computational capabilities to act as an independent peer. It both verifies the whole blockchain and relays other transactions. Obviously, with such IoT gateways, it will be possible to create a completely decentralized and trustless IoT infrastructure. Integration is easy because there is no special change required for end device communication.

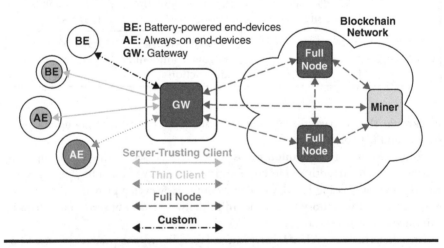

Figure 2.5 IoT–blockchain integration methods [74].

Gateway as a thin client: In this scenario, IoT gateways participate in the network as light nodes that store only certain parts of the blockchain (headers mostly) and request data from other full nodes, if needed. This integration model still depends on other full nodes on the system, so it is not possible to build a completely decentralized system just by using light client IoT gateways.

End devices as regular sensors: Battery-powered IoT end devices are very resource constrained in general, and it may be impossible to add a blockchain client. In this case, just like in any traditional LPWAN deployment, IoT gateways should be responsible to collect sensor data, package it, and push it to the blockchain.

End devices as server-trusting client: Historic programming interfaces like Bitcoin Client API (BCCAPI) provide a very simple interface for low-power IoT end devices. By using a similar, HTTP-based interface, it would be possible to push data directly into the blockchain without the need of computational resources.

End devices as thin client: In a scenario, where IoT end devices are always-on, it is possible to put a light client into the end devices. By having blockchain clients both in IoT end devices and in gateways, standardization in terms of data transmission and storage can easily be achieved.

2.7 IoT–Blockchain Evaluation

A blockchain-based IoT infrastructure will open up a lot of architectural possibilities which may help in securing and standardizing the IoT data flow and processing. In this section, three different IoT–blockchain infrastructure scenarios will be evaluated in terms of security.

2.7.1 Standardized Data Storage

By using a gateway-centric IoT infrastructure (like LoRa, Sigfox, or NB-IoT), it would be possible to store bulk data in a decentralized filesystem (like IPFS or Swarm) and keep the resource handles (i.e., file hashes) and associated metadata in an immutable blockchain. As a result of this approach, transmission and storage of IoT data may be standardized, therefore minimizing the attack surface.

As discussed, we used a gateway-centric technology for the IoT–blockchain integration. In the proposed system (Figure 2.6), a smart proxy is developed to store sensor data in Swarm and keep track of these files via a smart contract deployed in a private Ethereum network. This smart proxy is deployed in a LoRa gateway, and all the data coming from LoRa end devices are stored in Swarm and Ethereum [28].

In this design, standardization is forced at two important operations: how data is pushed out of the gateway and how data is accessed once it is stored in the backend. Fortunately, both of these operations are smart contract calls in JSON-RPC form. By adopting this approach, it is possible to streamline IoT software development with significant gains in terms of security.

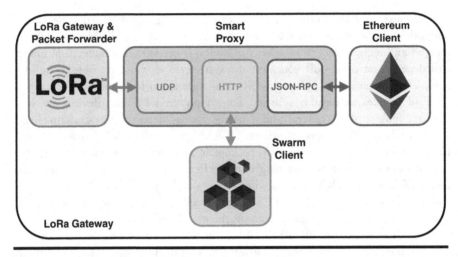

Figure 2.6 LoRa–Ethereum–Swarm solution architecture.

2.7.2 Decentralized Data Marketplace

Once a blockchain-based common data repository and querying interface is created, a blockchain-based marketplace for data sharing, exchanging, and trading may be established. In contrast to today's big data brokers like Google, Facebook, or Amazon, this structure will be a completely transparent and decentralized entity respecting user's preferences (e.g., strict opt-in) and using cryptographic techniques to ensure security and privacy for users.

Proposed decentralized data marketplace is implemented by using an Ethereum smart contract and Swarm filesystem [75]. Implementation is not only focused on interaction between IoT devices and Ethereum but also defining a clear interface for querying, browsing, and interacting with stored IoT data. Figure 2.7 shows the complete flow of the system. First, the device vendor registers itself to system via interacting with the smart contract (*vendor_register()*). Then, a device that belongs to that vendor is added (*add_valid_device()*). Added IoT device will upload encrypted sensor data to Swarm via HTTP protocol and receives a *file hash* representing that piece of information. After receiving the file hash, IoT device prepares *metadata* for that upload and pushes it to the smart contract via *sensor_data_push()* function.

On the other hand, any AI/ML provider may register to the system (*customer_register()*) and able to query sensor data using the collective metadata stored in the smart contract (*query_sensor()*). By using the *sensor_data_pull()* and *request_for_data()* functions, AI/ML providers may request and pay for a specific piece of data. In that call, AI/ML providers will share a public key and request the *content decryption key* for the purchased content. In the final step, IoT vendor will send

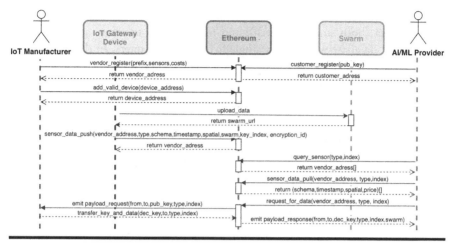

Figure 2.7   IoT–blockchain data market flow [75].

the *content decryption key* in encrypted form, later to be decrypted by the AI/ML provider (*transfer_key_and_data()*).

Proposed marketplace solution will provide both transparency and security due to end-to-end standardized flow for data producers and consumers.

2.7.3 Attack Profile Vault

In the previous sections, blockchain and decentralized filesystems are used for general data storage. However, it is preferable to use blockchain to store high-value data like virus profiles. By setting up honeypots and monitoring systems, security companies and researchers may be able to create manifests describing the attack pattern and mitigation strategies for any threat. Therefore, these threat manifests may be stored on blockchain thanks to its immutability. Being continuously accessed by IoT devices or gateways on the field, blockchain creates a decentralized, fast response mechanism against possible threats.

2.8 Conclusion

In this chapter, we searched the answers for enhancing IoT security by using blockchain technologies on device, network, and application levels. As presented, emerging technologies in IoT space like fog computing, LPWANs, and 5G enable blockchain technologies to be adopted at least at gateway (or base station) level. Therefore, it will be possible to facilitate a fault-tolerant, DDoS-resistant infrastructure with the help of new consensus protocols, decentralized filesystems, and decentralized application platforms. Such an infrastructure will standardize the way devices communicate with their backend, provide a unified way to interact with the data (both store and fetch), and may be used to create an immutable, global threat database which every single IoT device can query. In the end, due to its trustless and decentralized nature, it is evident that blockchain technologies will be extensively used in IoT and machine-to-machine transactions.

Acknowledgement

This work has been partially supported by Bogazici University Scientific Research Projects Fund (pr. no. 13500)

Figure 2.5. "IoT–blockchain Integration Methods" is copyrighted by IEEE. © 2019 IEEE. Reprinted, with permission, from "Work-in-progress: integrating low-power IoT devices to a blockchain-based infrastructure," 2017 International Conference on Embedded Software (EMSOFT).

Table 2.1. "LPWAN Connectivity Overview" is copyrighted by IEEE. © 2019 IEEE. Reprinted, with permission, from "Designing a Blockchain-based IoT with Ethereum, swarm, and LoRa: the software solution to create high availability with minimal security risks," IEEE Consumer Electronics Magazine (M-CE).

Figure 2.7. "IoT–blockchain Data Market Flow" is copyrighted by IEEE. © 2019 IEEE. Reprinted, with permission, from "IDMoB: IoT Data Marketplace on Blockchain," 2018 Crypto Valley Conference on Blockchain Technology (CVCBT).

References

1. IHS Tech, "Iot platforms: Enabling the internet of things," Feb. 2016. [Online]. Available: https://cdn.ihs.com/www/pdf/enabling-IOT.pdf.
2. IoT Analytics, "Iot analytics research," Feb. 2018. [Online]. Available: https://iot-analytics.com/state-of-the-iot-update-q1-q2-2018-number-of-iot-devices-now–7b.
3. S. Madakam, R. Ramaswamy, and S. Tripathi, "Internet of things (iot): A literature review," *Journal of Computer and Communications*, vol. 3, no. 05, p. 164, 2015.
4. C. Kolias, G. Kambourakis, A. Stavrou, and J. Voas, "Ddos in the iot: Mirai and other botnets," *Computer*, vol. 50, no. 7, pp. 80–84, 2017.
5. IoT Analytics, "The top 10 iot segments in 2018," Feb. 2018. [Online]. Available: https://iot-analytics.com/top-10-iot-segments-2018-real-iot-projects.
6. European Parliament, "General Data Protection Regulation (GDPR) (Regulation (EU) 2016/679)," Apr. 2016. [Online]. Available: http://eur-lex.europa.eu/legal-content/EN/TXT/PDF/?uri=CELEX:32016R0679&from=EN.
7. K. K. Patel, S. M. Patel et al., "Internet of things-IOT: Definition, characteristics, architecture, enabling technologies, application & future challenges," *International Journal of Engineering Science and Computing*, vol. 6, no. 5, 2016.
8. O. B. Sezer, E. Dogdu, M. Ozbayoglu, and A. Onal, "An extended iot framework with semantics, big data, and analytics," in *2016 IEEE International Conference on Big Data (Big Data)*. IEEE, 2016, pp. 1849–1856.

9. A. W. Services. "AWS IoT: IoT services for industrial, consumer, and commercial solutions," 2018. [Online]. Available: https://aws.amazon.com/iot.

10. J. Gubbi, R. Buyya, S. Marusic, and M. Palaniswami, "Internet of things (iot): A vision, architectural elements, and future directions," *Future Generation Computer Systems*, vol. 29, no. 7, pp. 1645–1660, 2013.

11. V. Pureswaran and P. Brody, "Device democracy: Saving the future of the internet of things," IBM Corporation, 2015.

12. A. Bahga and V. K. Madisetti, "Blockchain platform for industrial internet of things," *Journal of Software Engineering and Applications*, vol. 9, no. 10, p. 533, 2016.

13. IOTA Foundation, "Iota: a cryptocurrency for internet-of-things," Apr. 2016. [Online]. Available: https://www.iota.org.

14. MIT, "Neuromesh," Nov. 2016. [Online]. Available: www.atositchallenge.net/wp-content/uploads/2016/11/NeuromeshDescription.pdf.

15. Slock, "Slock.it," 2015. [Online]. Available: https://slock.it.

16. LO3 Energy, "Brooklyn Microgrid," 2018. [Online]. Available: https://lo3energy.com/innovations/.

17. S. Huh, S. Cho, and S. Kim, "Managing iot devices using blockchain platform," in *2017 19th International Conference on Advanced Communication Technology (ICACT)*. IEEE, 2017, pp. 464–467.

18. A. Dorri, S. S. Kanhere, R. Jurdak, and P. Gauravaram, "Blockchain for iot security and privacy: The case study of a smart home," in *2017 IEEE International Conference on Pervasive Computing and Communications Workshops (PerCom Workshops)*. IEEE, 2017, pp. 618–623.

19. M. S. Ali, K. Dolui, and F. Antonelli, "Iot data privacy via blockchains and ipfs," in *Proceedings of the Seventh International Conference on the Internet of Things*. ACM, 2017, p. 14.

20. B. Lee and J.-H. Lee, "Blockchain-based secure firmware update for embedded devices in an internet of things environment," *The Journal of Supercomputing*, vol. 73, no. 3, pp. 1152–1167, 2017.

21. M. Baza, M. Nabil, N. Lasla, K. Fidan, M. Mahmoud, and M. Abdallah, "Blockchain-based firmware update scheme tailored for autonomous vehicles," arXiv preprint arXiv:1811.05905, 2018.

22. W. Shi, J. Cao, Q. Zhang, Y. Li, and L. Xu, "Edge computing: Vision and challenges," *IEEE Internet of Things Journal*, vol. 3, no. 5, pp. 637–646, 2016.

23. K. Mekki, E. Bajic, F. Chaxel, and F. Meyer, "A comparative study of lpwan technologies for large-scale iot deployment," ICT Express, 2018.

24. R. S. Sinha, Y. Wei, and S.-H. Hwang, "A survey on lpwa technology: Lora and nb-iot," *Ict Express*, vol. 3, no. 1, pp. 14–21, 2017.

25. 3GPP, "Progress on 3GPP IoT," 2016. [Online]. Available: www.3gpp.org/news-events/3gpp-news/1766-iotprogress.

26. J. Bardyn, T. Melly, O. Seller, and N. Sornin, "IoT: The era of LPWAN is starting now," in *European Solid-State Circuits Conference, ESSCIRC Conference 2016: 42nd*. IEEE, 2016, pp. 25–30.

27. Nokia, "LTE evolution for IoT connectivity," 2015. [Online]. Available: http://resources.alcatel-lucent.com/asset/200178.

28. K. R. Ozyilmaz and A. Yurdakul, "Designing a blockchain-based iot with ethereum, swarm, and lora: The software solution to create high availability with minimal security risks," *IEEE Consumer Electronics Magazine*, vol. 8, no. 2, pp. 28–34, 2019.

29. S. Li, L. Da Xu, and S. Zhao, "5g internet of things: A survey," *Journal of Industrial Information Integration*, vol. 10, pp. 1–9, 2018.

30. R. T. Tiburski, L. A. Amaral, and F. Hessel, "Security challenges in 5g-based iot middleware systems," in *Internet of Things (IoT) in 5G Mobile Technologies*. Springer, 2016, pp. 399–418.

31. McKinsey, "Top10 insights of 2018," 2018. [Online]. Available: www.mckinsey.com/about-us/new-at-mckinsey-blog/top-10-insights-of-2018.

32. Y. LeCun, Y. Bengio, and G. Hinton, "Deep learning," *Nature*, vol. 521, no. 7553, p. 436, 2015.

33. A. Javaid, Q. Niyaz, W. Sun, and M. Alam, "A deep learning approach for network intrusion detection system," in *Proceedings of the 9th EAI International Conference on Bio-inspired Information and Communications Technologies (formerly BIONETICS)*. ICST (Institute for Computer Sciences, Social-Informatics and Telecommunications Engineering, 2016, pp. 21–26.

34. M. Barreno, B. Nelson, A. D. Joseph, and J. D. Tygar, "The security of machine learning," *Machine Learning*, vol. 81, no. 2, pp. 121–148, 2010.

35. A. L. Buczak and E. Guven, "A survey of data mining and machine learning methods for cyber security intrusion detection," *IEEE Communications Surveys & Tutorials*, vol. 18, no. 2, pp. 1153–1176, 2016.

36. W. Diffie and M. Hellman, "New directions in cryptography," *IEEE Transactions on Information Theory*, vol. 22, no. 6, pp. 644–654, 1976.

37. R. L. Rivest, A. Shamir, and L. Adleman, "A method for obtaining digital signatures and public-key cryptosystems," *Communications of the ACM*, vol. 21, no. 2, pp. 120–126, 1978.

38. L. Lamport, "Constructing digital signatures from a one-way function," Technical Report CSL-98, SRI International Palo Alto, Tech. Rep., 1979.

39. R. Merkle, "Secrecy, authentication, and public key systems," Ph. D. Thesis, Stanford University, 1979.

40. N. Koblitz, "Elliptic curve cryptosystems," *Mathematics of Computation*, vol. 48, no. 177, pp. 203–209, 1987.

41. V. S. Miller, "Use of elliptic curves in cryptography," in *Conference on the Theory and Application of Cryptographic Techniques*. Springer, 1985, pp. 417–426.

42. S. Nakamoto, "Bitcoin: A peer-to-peer electronic cash system," 2008. [Online]. Available: https://bitcoin.org/bitcoin.pdf.

43. R. C. Merkle, "A digital signature based on a conventional encryption function," in *Conference on the Theory and Application of Cryptographic Techniques*. Springer, 1987, pp. 369–378.

44. S. Popov, "The tangle," cit. on, p. 131, 2016.

45. S. King and S. Nadal, "Ppcoin: Peer-to-peer crypto-currency with proof-of-stake," Self-Published Paper, August, vol. 19, 2012.

46. V. Buterin and V. Griffith, "Casper the friendly finality gadget," arXiv preprint arXiv:1710.09437, 2017.

47. D. Larimer, "Delegated Proof-of-Stake (DPoS)," Bitshare Whitepaper, 2014.

48. Lisk, "Lisk consensus algorithm," Apr. 2019. [Online]. Available: https://lisk.io/documentation/lisk-protocol/consensus.

49. EOSIO, "Eos.io technical white paper v2," Mar. 2018. [Online]. Available: https://github.com/EOSIO/Documentation/blob/master/TechnicalWhitePaper.md.

50. Steem, "Steem: An incentivized, blockchain-based, public content platform," Mar. 2018. [Online]. Available: https://steem.com/steem-whitepaper.pdf.

51. M. Castro, B. Liskov et al., "Practical byzantine fault tolerance," in *OSDI*, vol. 99, 1999, pp. 173–186.

52. G. Bracha, "Asynchronous byzantine agreement protocols," *Information and Computation*, vol. 75, no. 2, pp. 130–143, 1987.

53. M. Vukolić, "The quest for scalable blockchain fabric: Proof-of-work vs. bft replication," in *International Workshop on Open Problems in Network Security*. Springer, 2015, pp. 112–125.

54. E. Androulaki, A. Barger, V. Bortnikov, C. Cachin, K. Christidis, A. De Caro, D. Enyeart, C. Ferris, G. Laventman, Y. Manevich et al., "Hyperledger fabric: A distributed operating system for permissioned blockchains," in *Proceedings of the Thirteenth EuroSys Conference*. ACM, 2018, p. 30.

55. Rinkeby, "Rinkeby: Ethereum testnet," Apr. 2017. [Online]. Available: www.rinkeby.io/.

56. Hyperledger Sawtooth, "Proof of elapsed time (poet)," 2018. [Online]. Available: https://sawtooth.hyperledger.org/docs/core/releases/1.0/architecture/poet.html.

57. J. Götzfried, M. Eckert, S. Schinzel, and T. Müller, "Cache attacks on intel sgx," in *Proceedings of the 10th European Workshop on Systems Security*. ACM, 2017, p. 2.

58. J. Van Bulck, M. Minkin, O. Weisse, D. Genkin, B. Kasikci, F. Piessens, M. Silberstein, T. F. Wenisch, Y. Yarom, and R. Strackx, "Foreshadow: Extracting the keys to the intel {SGX} kingdom with transient out-of-order execution," in *27th {USENIX} Security Symposium ({USENIX} Security 18)*, 2018, pp. 991–1008.

59. T. Rocket, "Snowflake to avalanche: A novel metastable consensus protocol family for cryptocurrencies," 2018. [Online]. Available: https://avalanchelabs.org/avalanche.pdf.

60. S. Goldwasser, S. Micali, and C. Rackoff, "The knowledge complexity of interactive proof systems," *SIAM Journal on Computing*, vol. 18, no. 1, pp. 186–208, 1989.

61. E. B. Sasson, A. Chiesa, C. Garman, M. Green, I. Miers, E. Tromer, and M. Virza, "Zerocash: Decentralized anonymous payments from bitcoin," in *Security and Privacy (SP), 2014 IEEE Symposium on. IEEE*, 2014, pp. 459–474.

62. B. Bünz, J. Bootle, D. Boneh, A. Poelstra, P. Wuille, and G. Maxwell, "Bulletproofs: Short proofs for confidential transactions and more," in *2018 IEEE Symposium on Security and Privacy (SP)*. IEEE, 2018, pp. 315–334.

63. T. E. Jedusor, "Mimblewimble," 2016.

64. D. Williamson, "The aztec protocol," 2018. https://github.com/AztecProtocol/AZTEC/blob/master/AZTEC.pdf.

65. G. Maxwell, "Confidential transactions," 2015. https://people.xiph.org/greg/confidentialvalues.txt.

66. J. K. Liu, V. K. Wei, and D. S. Wong, "Linkable spontaneous anonymous group signature for ad hoc groups," in *Australasian Conference on Information Security and Privacy*. Springer, 2004, pp. 325–335.

67. S. Noether, "Ring signature confidential transactions for monero," IACR Cryptology ePrint Archive, vol. 2015, p. 1098, 2015.

68. J. Benet, "Ipfs-content addressed, versioned, p2p file system," arXiv preprint arXiv:1407.3561, 2014.

69. J. H. Hartman, I. Murdock, and T. Spalink, "The swarm scalable storage system," in Distributed Computing Systems, 1999. *Proceedings. 19th IEEE International Conference on*. IEEE, 1999, pp. 74–81.

70. S. Wilkinson, T. Boshevski, J. Brandoff, and V. Buterin, "Storj a peer-to-peer cloud storage network," 2014.

71. S. Prime, "Solidity security: Comprehensive list of known attack vectors and common anti-patterns," Oct. 2018. [Online]. Available: https://blog.sigmaprime.io/solidity-security.html.

72. M. A. Khan and K. Salah, "Iot security: Review, blockchain solutions, and open challenges," *Future Generation Computer Systems*, vol. 82, pp. 395–411, 2018.

73. S. B. Venkatakrishnan, G. Fanti, and P. Viswanath, "Dandelion: Redesigning the bitcoin network for anonymity," arXiv preprint arXiv:1701.04439, 2017.

74. K. R. Özyılmaz and A. Yurdakul, "Integrating low-power IoT devices to a blockchain-based infrastructure: work-in-progress," in *Proceedings of the Thirteenth ACM International Conference on Embedded Software 2017 Companion*. ACM, 2017, p. 13.

75. K. R. Özyilmaz, M. Doğan, and A. Yurdakul, "IDMoB: IoT data marketplace on blockchain," in *2018 Crypto Valley Conference on Blockchain Technology (CVCBT)*. IEEE, 2018, pp. 11–19.

Chapter 3

An Overview of the Dark Web

Shahrin Sadik
International Islamic University of Chittagong

Mohiuddin Ahmed
Edith Cowan University

Contents

3.1 Introduction

The Internet has been one of the greatest blessings of science and technology for the modern world. It has constantly worked as a motivation for bringing out regular changes according to the needs of the developing era. It has several significant uses which boost the regularities of casual life, and it has only made them applicable because of the existence of its decentralized nature [1]. The Internet has connected the globe by providing its wide feature of connectivity and has contributed to the world with its concept of traveling longer distances with higher speed in a shorter amount of time. Signals used here for traveling have an equivalent speed of light, which makes it more convenient and reliable. Thus, it has been easier for the virtual world to cover up all the possible and impossible geographical distances than in the real world. Distance is no longer a barrier for global connectivity. This specific feature of the Internet, though, promotes a high level of interactivity, whereas it somehow also increases the risk of misusing the virtual world for malicious intentions. Basically, there are three layers of Internet as shown in Figure 3.1. The surface web, deep web, and dark web are the three levels of the Internet, which provide three different contents for three respective users.

The surface web is the one which is easily available in the search engine for every user of the world. The deep web is a little more crucial part of the Internet as this is limited to only a few members of the organizations by allowing them to have their own user identity as per the requirements of the website. It is usually hidden from the regular users and only available for these limited users. The third layer of the Internet is the dark web, which is the most dangerous part since it deals with all the unethical and inhuman activities or illegal businesses. The dark web has been the biggest threat to the modern world; it is not easily accessible by everyone using normal search engines, but rather only by those specific users who deal with illegal acts and remain anonymous for the world. This chapter plans to summarize the concept of the dark web. Sections 1.1–1.4 direct the reader to the details of the dark

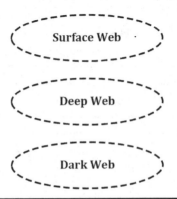

Figure 3.1 Layers of the Internet.

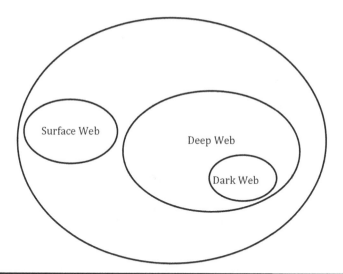

Figure 3.2   Internet and layered web.

web in different layers as shown in Figure 3.2 and security concern. Section 3.2 aids the reader to know about the taxonomy of the World Wide Web (WWW). Sections 3.3 and 3.4 cover the governance and the future of the dark web. The later part, i.e., Section 3.5, concludes this chapter.

3.1.1 What Is the Dark Web?

The dark web is a collection of websites which are not available on general search engines and prevail only in certain encrypted networks. A tool named Tor encryption is used to hide the user identity in almost all the sites of the dark web. This tool aids the user to hide his or her identity and restricts the correct location from being shared. It also acts as a shield for all the activities performed in the dark web. It is necessary for the users of the dark web to use Tor to hide their IP addresses. It has to pass through several layers of encryption to allow the hidden feature to be used by the users of the dark web. It is possible for anyone to visit dark web using Tor, but one has to make sure that the identity is hidden so that they do not get into the trap of the backend users who are extremely dangerous.

The dark web is referred as the furthest corner of the deep web where all the information is deliberately concealed to not make it publicly available through traditional search engines [2]. It has rapidly become a pathway for illegitimate purposes and illicit activities. This has attracted great attention of the policy and law makers because of its constant exploitation. The Silk Road was one of the main dark websites which was used to promote several illegal and malicious activities and also served as one of the biggest platforms for online illicit shopping.

Many statistics show that the vendors of this specific platform were scattered in more than 10 countries and supplied illegal items to around 100,000 buyers. This site made an overall profit of almost $1.2 billion within two to three years by selling their products. Later, the federal agents demolished this website and saved both the virtual and real world from being oppressed. The question of having enough tools to combat the new virtual underworld has drawn the attention of law enforcements and other relevant officials to solve this mysterious issue of whether or not we are all well prepared.

3.1.2 How Is the Dark Web Used?

There is no proper estimation of how much of the deep web is used specifically for illegal or legitimate arena of the dark web. The Onion Router [3] most commonly known as Tor and the Invisible Internet Project abbreviated as I2P along with many other anonymous and decentralized nodes of the network are used to access the dark web. This was originally created and released by the USA Naval Research Laboratory for being able to communicate online keeping the user's identity hidden. Tor generally refers to both the network which operates all the Tor connections and the software which is installed to use and run Tor in the system. The users of Tor basically do not create a direct connection and rather pass through a series of tunnels which are virtually integrated in it. This allows sharing information in the website without compromising the privacy of any user. There is a three-way pass for the Tor users, and the last one is referred as "the exit relay." The web traffic is routed through the other users' system by the authentic user to avoid the identification and tracking of the real user. Tor particularly creates three imaginary layers just as the natural onion has, and it routes the movement in those layers in order to hide the original identity of the users by hiding the IP address.

3.1.3 Where Is the Dark Web Used?

The dark web is the third layer of the Internet which majorly deals with all the illegal activities and is kept hidden from the traditional web browsers. It uses an anonymous network; popularly, Tor is used for browsing and to get connected with its users. It is usually composed of several encrypted networks which generally hide the user's identity and location so that the real information of the user remains concealed. These websites are a minor portion of the deep web and have a limited way of access; i.e., these sites can be accessed only by some specialized software. Tor allows the users to access the pages referred as "onions" which are encrypted for extreme security and privacy. Here, the requests for the connections are routed several times before reaching their assigned destination. Tor particularly also aids in bypassing state censorship in order to protect the user's privacy; i.e., it eventually supports the officials and the law enforcement agencies [4].

The dark web is considered to be a specialized digital underworld which specially promotes all the illicit activities such as trading pornography, buying and selling of drugs and weapons, hacking software and users, spoiling the bank or any online account detail, online human trafficking, uploading and watching videos of humans torturing animals, buying and selling of human meat, and even contractual murder [5]. The anonymity feature of the dark web is the only reason of its growing demand for both legitimate and illicit procedures which are involved for such online activities.

Despite this, the dark web is also used by many security organizations for some valid and logical reasons of privacy. For example, in this technological era, some of the recent journalists intend to use the dark web with the purpose of concealing their identity and location. This is done so as to hide their whereabouts and ask relevant questions to the interviewer for their upcoming and trending updates. They utilize the advantage of anonymity and even use the dark web to store and secure their significant documents. The dark web is not only used for illegitimate purposes but also facilitates the privacy of a user by aiding him or her to hide what he or she has been doing in a certain session.

3.1.4 Why Is It Important for Security?

The anonymity feature of the dark web not only aids criminals but also helps government agencies and law enforcement teams to conduct online surveillance of illicit activities. It also allows the Federal Bureau of Investigation (FBI) and law agencies to keep a track of all the malicious activities happening online. The FBI is trying to find ways to unmask the anonymous users of Tor. This is specially planned and targeted to identify online sexual predators, pedophiles, hackers, illegal buyers and sellers, extortionists, etc. It equally supports the military and intelligence to follow up the environment and study about it so as to prevent terrorist groups from planning and plotting against the nation. The military and intelligence bureaus are also able to use different tactics to spoil the plans of terrorists and deceive them online. This ensures the security of significant information present online and tends to protect the nation from being targeted for severe attacks online or offline.

3.2 Taxonomy of the Web

The WWW is comprised of a clear net, which is also known as the surface web, and the deep web. The surface web is the one which only covers 4% of the traditional web (WWW), whereas the rest (96%) deals with the deep web. Google is traditionally one of the largest search engines, but its indexing is only limited to 6%–7% of the whole web. Ninety-six percent of the total web is not easily accessible by standard and easily available search engines because these contents are not indexed in it. Web crawl or Spider web is being used by these search engines, which promotes the procedure of moving from one hyperlink to the other carrying necessary information and preserves the same index. At times, few authors also index their websites with a motive to make them available in the results of the other search engines. It was observed in a study of the year 2016 that there are around 46% of the overall world population who are available and globally connected on the Internet, but there are no apparent data on traceability of about half of this web-based population. According to another recent study in 2019 on the total number of websites present on the Internet, there are approximately 1.9 billion websites in total. But as the number of users and websites is consistently fluctuating, there is no proper estimation of how many sites are being developed and used on a daily basis.

3.2.1 Clear Net or the Surface Web

The surface web is considered to be the publicly available Internet, which has no limited access or restrictions for the worldwide users. This part is not accountable for any payment or authentication from the users for accessing this web and makes it reliable as it provides the identifying features of both the host and the user in accordance with the law. For example, there are many websites that offer or publish several real-time updates or regular updates in a timely manner, which in turn are accountable only for publishing their contents. Along with this, the users surfing and using these websites are also accountable for visiting the relevant sites, and their visit is recorded and observed by the server and the one who controls and monitors the network, respectively. It also keeps a track of all the IP addresses visiting the sites and their contents. According to a major analysis of one of the science magazines in 1998, until 1998 there were about 320 million documents present in the surface web. This further increased to 800 million as per the study in 1999 and later to a size of around 18,700 gigabyte on the basis of HTML. A recent 21st-century study shows that there has been a huge growth of documents, i.e., almost adding up to around 7.5 million per day which eventually summed up to about 2.5 billion documents in the surface web in total.

3.2.2 Deep Web

The deep web consists of some crucial information such as online banking or payment system, sites relevant to user login and passwords, access to limited contents, and all the scripted and the non-HTML contents [6]. Along with this, there are also contents which can only be accessed through specific websites. It is quite impossible to evaluate the actual size of the deep web because it conceals a large number of the databases involved in it. Such contents are apt to only relevant information and the related domain. It further relinquishes the search results and also restricts its use for general users. It is an astonishing fact that the deep web is much more critical and faces huge traffic because of its wide use. It is linked very crucially and complexly and anticipates related and reliable information [6]. It is also not possible for the contents to be indexed by any search engine because most of the data are password-protected and have restricted and limited access only by the software selected for the purpose [7]. It basically forms a separate portion of an internal network. For example, if there is an online bank transaction of a user, it is only restricted to the individual as it is private, and he or she does not allow anyone else to access the same. This is one of the simplest examples of the deep web as it limits the sharing of contents to a certain individual as per the requirement. This makes both the bank and the client identifiable to each other, and any one of these parties can easily keep a track of the other if and when required [8]. The data present in the deep web is generally dynamic, and it is not even related or linked to the other webpages of the same site. It is nearly impossible to evaluate the size of the deep web as it is expected that the size of the deep web is around 4,000 to 5,000 times larger than that of the surface web [9].

3.2.3 Dark Web

The dark web, as already explained, is a minor fragment of the deep web which lets the user be anonymously active throughout his/her each and every Internet session. This website is the farthest part of the deep web, which is crucially the largest portion of the WWW; i.e., it almost uses about 96% of the WWW. The dark web eventually directs and somehow promotes illicit activities, but it is also being used for various significant tasks by law enforcement agencies. This is commonly termed as "the digital underworld" as it has been widely taken over by the anonymous users who have the ultimate power of being hidden and traffic illegal items. The dark web has always been a mystery for the world from the very beginning and ending of its largest website "The Silk Road." It was identified and was later shut down following many laws and orders by the officials and law enforcing agencies. It had a maximum contribution toward the online illicit business or marketplace. This dealt with all the illegitimate issues around the world.

3.3 Governance of the Dark Web

Due to the increasing amount of decentralized communication, it is becoming difficult to regulate. This may lead to extreme sharing of unethical contents across the Internet, and it will ultimately be hard to keep these situations under control and in the watch list of the law. The main center of attraction in the current phase of technological advancement is to somehow be able to deanonymize all the secret and hidden services provided online through this dark web. It is to make sure that all the facilities for conducting anonymous crimes are stopped from being proliferated. To put a halt on all the illegal activities of the dark web, FBI and the related law enforcement agencies have been working to take severe actions against this illicit business. They have also discovered fruitful outcomes in the path of this digital underworld marketplace and were able to successfully shut down the 2.0 version of The Silk Road and the Europol in the years 2013 and 2014, respectively. Yet, the 3.0 version of The Silk Road is still available online. The FBI and the legal agencies have appointed several white hat hackers to keep a track of all the illegal and criminal activities as well as the people related to them. These law agencies are also trying to be more technologically advanced so as to prevent the criminals or the dark web users from conspiring against humankind. They are also moving with the motive to follow and keep track of the digital transactions with Bitcoin and be aware of any sort of online money laundering involved.

3.4 Future of the Dark Web

It is predicted that the dark web will become darker and deeper with the increasing technological aspects. The dark web is expected to be even more customer-friendly and innovative, difficult to censor, and more decentralized. It is also anticipated that it will become difficult for the law enforcement agencies to detect the dark web users because of the high risk of their increasing anonymity. This digital marketplace will eventually develop with the growing demand of the customers, and the private delivery of its items will be restricted to some anonymous delivery, which will ensure the secrecy of the traceability of the relevant delivery. There might be several situations in the future where an online judiciary system for the dark web will be needed so as to handle criminal activities online as there are no proper offline laws and orders to handle these delicate issues. The dark web will eventually become way safer and secure than the way in which it currently exists. Online transactions through Bitcoin will rapidly increase and the implementation of anonymity will also proportionally increase, which will in turn ultimately make it hard to trace such Bitcoin transactions.

3.5 Conclusions

The mysterious web which is usually referred to as the "dark web" or the "dark net" is a place where some questionable activities take place. It is usually invisible in the index of traditional search engines, which makes it impossible for the casual users to track or find out the whereabouts of a dark web user. The only thing which is available using the traditional method is the surface web. Though the surface web is used by several active users in the present time, the dark web is restricted to only a small number of people. Most of the users of the dark web are involved in various illicit and criminal activities. Money laundering through the dark net or dark web is something very common nowadays along with other cyber-crimes, but it is not only limited to illegal activities [10]. It also has some very usual benefits, such as the WikiLeaks, which support the appropriate agencies respectively, and along with this, it also allows the Chinese users to access few of the sites whose use is restricted in their locality. This dark web uses a sophisticated encryption technique which provides three layers of encryption in order to hide the user's identity and location, or in other words, it helps to alter the original IP address being used. However, it is a matter of great importance in recent times as it certainly deals with the majority of the digital underworld activities and its usage is increasing exponentially. Thus, law enforcement agencies are looking out for some unorthodox ways to keep a track of these anonymous criminals so as to make the dark web more useful rather than being perilous for everyone.

References

1. What is the Dark Web & How to Access it, accessed from: www.techadvisor.co.uk/how-to/internet/dark-web-3593569/.
2. Michael Chertoff and Toby Simon, "The Internet of the Dark Web on Internet Governance and Cyber Security," Global Commission on Internet Governance, Paper Series: No. 6, February 2015, p. 1.
3. Roger Dingledine, Nick Mathewson, and Paul Syverson, "Tor: The Second-Generation Onion Router", *Proceedings of the 13th USENIX security Symposium*, San Diego, August 2004.
4. Mysteries of Dark Web, accessed from: www.cnbc.com/2018/09/06/beyond-the-valley-understanding-the-mysteries-of-the-dark-web.htmls.
5. What Is the Dark Web and Why Do People Use It? Accessed from: www.lifewire.com/access-the-dark-web-3481559.
6. M. Bergman, Presented by Mat Kelly, CS895 – Web-based Information Retrieval, The Deep Web: Surfacing Hidden Value, Old Dominion University, September 27, 2011.
7. Vincenzo Ciancaglini, Marco Balduzzi, Robert McArdle, and Martin Rösler, "Trend Micro, A Trend Labs," Research paper Below The Surface: Exploring The Deep Web, Japan, 2015.
8. Iflah Naseem, Ashirr K. Kashyaap, and Dheeraj Mandloi, "Exploring Anonymous Depths of Invisible Web and the Digi-Underworld," IJCA, 2016.
9. Kristin Finklea, "Dark Web," Congressional Research Service, March 2017.
10. Immunity on the Dark Web as a Result of Blockchain Technology, accessed from: https://codeburst.io/immunity-on-the-dark-web-as-a-result-of-blockchain-technology-6693eb087bdd.

CYBER
GOVERNANCE

Chapter 4

Cyber Security in the Public Sector: Awareness of Potential Risks among Public Policy Executives

---

Tahmina Rashid
University of Canberra

Khalid Chauhan
Govt of Azad Jammu Kashmir

Contents

4.1 Introduction

Cyber security of online users of internet is gaining currency as both a global and a local issue; though locally in Pakistan, not much attention has been paid to the threats emanating from the cyberspace. Globally, the term "cyber security" has been defined by the International Standards Organization (ISO) 27032 as the "preservation of confidentiality, integrity and availability of information in the Cyberspace" [1]. Similarly, "cyberspace" is usually referred to as "the complex environment resulting from the interaction of people, software and services on the Internet by means of technology devices and networks connected to it, which does not exist in any physical form." Hundley and Anderson [2] note that "cyberspace safety" results from the inherent unpredictability of computers that can potentially threat the physical and human environments; these threats arise from software as well as hardware failures and cannot be corrected through technology solutions [3]. Both the terms "cyber security" and "cyberspace" have gained more relevance and significance in the last few years. Until the end of the 20th century, little was known about cyber-attacks, cybercrimes, cyber-warfare, and cyber harassment [4]. Similarly, the terms such as "malware," "worm," "virus," "hacking," "data theft," "identity theft," "griefing," "spoofing," "spyware," and "phishing scams" were almost nonexistent [5,6]. This new jargon and these new threats have implications about the way individuals and states conduct their daily business [7]. Nye aptly notes cyber power in the current context and defines it as:

> [A] set of resources that relate to the creation, control, and communication of electronic and computer-based information-infrastructure, networks, software, human skills. This includes not only the Internet of networked computers, but also Intranets, cellular technologies, and space-based communications. Defined behaviorally, cyber power is the ability to obtain preferred outcomes through use of the electronically interconnected information resources of the cyber domain. Cyber power can be used to produce preferred outcomes within cyberspace, or it can use cyber instruments to produce preferred outcomes in other domains outside cyberspace [8].

Cyber power and cyber security are vital for the physical security of the countries, or even more. Individuals, organizations, and even nation states can potentially exploit the cyberspace for intrusion into the systems and processes for inflicting harm, espionage, and even disruption of services. In the United States, there is an ongoing debate about the alleged Russian interference into the Democratic National Convention (DNC) server, manipulating data to influence the outcome of the presidential elections [9]. This suggests the vulnerability and the potential exploitation of the unguarded systems or a lack of awareness of how to move in the relatively unfamiliar, unchartered terrain of cyberspace [10]. While the international consensus on cyber security parameters is yet to emerge, countries like the UK have strengthened their institutional mechanisms, processes, and protocols to thwart the possible threats. UK launched its National Cyber Security Strategy to defend, deter, and develop skills to tackle the challenges of cyber security [11].

With the increase in the number of online users of internet in Pakistan, the probability of exposure to the cyber-attack is more than ever before. Increasingly, the state business is being conducted through the use of internet, and there is an increased risk to cyber security of confidential information for policy makers and public offices. Pakistan has promulgated Prevention of Electronic Crimes ACT 2016, which deals with the breaches of cyber security [12]. However, less is known about how aware, prepared, and protected the public sector executives are from the threats emanating from cyberspace. In order to bridge the gap, this chapter provides a context of cyber security threats through a snapshot of some major recent incidents from around the world. The basic premise is that the public sector executives of Pakistan are not immune to these threats and must prepare for any eventuality, hence the need of awareness of threats and preparation to that effect. It is also a reminder that it is not merely a security threat (as was experienced during the Stuxnet), but a common threat that can potentially affect all public services as well as people [13]. It then presents the results of the survey of the case study on cyber security awareness among senior policy makers in Pakistan.

4.2 Statement of the Problem

Cyber security remains an issue even in the developed countries, while the Global South has the added challenges due to the lack of awareness among internet users. "Internet World Stats" indicates that in 2018, there were 4,208,571,287 internet users in the world and 44,608,065 internet users in Pakistan; this reflects a whopping increase in the number of internet users in Pakistan since 2000, reflecting an increase in internet penetration from 0.7% to 17.8% of the population. Although Pakistan is making efforts to improve cyber security by regulating the sector through appropriate policies and has passed legislation, Prevention of Electronic Crimes Act 2016 to deal with the breaches of cyber security, there is very little done at the organizational level to have cyber security policies in place, in line with the legislation [12]. This makes Pakistan a good case study which analyzes how public offices are operating their systems using the email servers that reside in other countries putting the data and organizations at serious cyber security risks. This case study will explore the level of awareness among public sector executives of threats emanating from the cyberspace through the use of internet, especially because senior executives are not digital natives yet play a significant role in public policy making and handling the potential cyber security threats. It would particularly investigate how aware, prepared, and protected are senior policy executives in Pakistan about the possible threats and attacks emanating from the cyber space?

4.3 Significance and Scope of the Case Study

This case study is important because it comes at a time when the number of online users of internet has increased manifold in Pakistan, increasing vulnerability to threats emanating from the cyberspace. However, little research is available on the awareness aspect of cyber threats and vulnerability, hence the significance of this research. Prevention of Electronic Crimes Act 2016 in Pakistan is intended to tackle the reported breaches of the cyber security and the harm emerging from the actions of the cyber criminals. Prevention in part is a function of the awareness of the malaise. It is important that the extent of awareness of cyber security issues is known so that adequate measures could be taken.

4.4 Review of the Literature

In the UK, Furnell et al. conducted a survey of 415 home users of internet to gauge their awareness and attitudes toward security issues and safeguards [14]. They found lack of awareness and knowledge in the novice as well as advanced users of internet. While the novice users had little knowledge about how to protect themselves, the advanced users who claimed to be knowledgeable could not demonstrate safe practices. They concluded that "awareness of official and mass media efforts to educate the population can be shown to be lacking in engagement and impact." They recommended "research into new models of engagement and awareness promotion that will go beyond simple definitions and shallow knowledge to provide a more effective learning foundation" for the people so they could protect themselves from the cyber threats. A similar survey of major Australian businesses noted lack of awareness and recognition about cyber-attacks, and recommended the need for a Risk Register and accessible and secure method for individuals to report, recognize, and avoid common types of cyber crimes [15]. Whitty et al. examined the cyber security behaviors in terms of individual differences, to explore the likelihood of sharing passwords, and found that despite the public awareness campaigns, individuals still indulge in risky online behavior like sharing passwords [16]. In their study, they found that 51.1% of the respondents in the past had shared their passwords, and the percentage was higher among young users. They believe that it can be attributed to the relatively higher preference for online communication among younger people due to access to technologies, and being digital native, they prefer online engagement, friendships, photo sharing as well as online banking. They recommend that awareness campaigns must engage young people for raising awareness about online safety and cyber security threats. Similar studies in the Asian context recommend assessing cyber security awareness, issues, and challenges to design appropriate policy frameworks [17]. There is very little research in the context of Pakistan regarding cyber security issues and awareness; hence, the literature from other contexts provides a context for this case study research.

4.5 Method of Study

This research uses the mix method approach and examines the broader issues and challenges of cyber security summarizing some incidents where cyber security of institutions and countries was breached. This research expands upon cyber security issues and challenges, explores some of the recent cyber security incidents in the world to emphasize the significance of the need for awareness of the threats emanating from cyberspace, and finally discusses the results of the survey conducted. It then narrows down the focus to pay closer attention to the case study context. The choice of method is in line with the view that the reality is far more complex from the arbitrary qualitative-quantitative divide practiced in the diametrically opposite epistemological paradigms of positivism and interpretivism. The data for the case study was collected from 54 senior public sector executives using a structured questionnaire survey. Data collected was analyzed by organizing it around key questions. The key findings are then presented under various key themes that emerged from the data. Finally, the research makes some recommendations based on the analysis, which can be potentially replicated in similar contexts.

4.6 Cyber Security—Issues & Challenges

4.6.1 What is Cyber Security?

The origin of the term "cyber security" dates back to 1988, when the first ever computer virus, Morris worm, infected sixty thousand internet-connected computers making them unusable [18]. Now cyber security is considered as the protection against attacks from outside computers aimed at distorting websites, computer systems, and the internet-based financial services, which are essential for the daily functioning of the society [4,19]. As the government organizations, markets, educational institutions, hospitals, and even airports have increasingly become dependent on the internet to conduct their daily business [20], for the storage and retrieval of personal data and information, individuals and governments are more susceptible to cyber-attacks [13,21]. It is estimated that the industry and the businesses are suffering more than 400 billion dollars of loss due to cybercrimes [22,23]. Among the victims are individuals, financial institutions, entertainment companies, corporations, as well as small and medium businesses. For governments, the magnitude of threat is even more serious and includes national security and impact on the functioning and service delivery by various ministries. Hacking is a new tactic of warfare as well as a business and a hobby of geeks [24]. Since the internet defies conventional physical boundaries, the cyber criminals are known to have acted from countries such as Russia, China, Nigeria, Vietnam, and Brazil, known generally as the global hotspots of cybercrime [25]. While at the one end of the continuum of cyber security threats are the hackers, at the other end are the unprotected computer systems and software, unaware employees [15,26], and lack of procedures making the possibility of cyber threats imminent and real [27]. This warrants holistic approaches to the understanding of the concepts of cyber security. Whereas the overarching concern in the field of cyber security is the protection of computer systems, the same is intended to be done through the identification and addressing of potential vulnerabilities in three domains, namely, (1) people, (2) processes, and (3) technologies.

4.6.2 People, Processes, and Technology

People as the online users of internet are most vulnerable to cyber security attacks. Hackers use emails with infected links to launch social engineering attacks on the individuals using internet independently or as part of global and national hackers groups [13]. Consequently, unaware individuals clicking the malicious links can result in theft of personal and organizational data such as credit card detail or the employee information.

Processes and policies define the extent to which organizations and individuals are protected and prepared to tackle cyber security threats. Individuals need to know the best processes and practices, whereas organizations need to have best possible requisite policies and rules, the do's and don'ts of using the cyberspace. Simple policies on how to ensure confidentiality of official communication and confidential data, that is, having a strong password to access office emails and computers, can save individuals and organizations from potential cyber security attacks [4].

Technology plays an important role in cyber security. For example, antivirus software packages, firewalls, detection devices, and cameras are used to enhance cyber security. As the technology is continually changing, upgrading software packages and systems largely depends on organizations and individuals to ensure the security of the use of technology, as technology cannot protect individuals and organizations on its own.

Any neglect in any of these three domains can potentially be exploited for cyber-attacks, namely, theft of data, access to emails accounts, and Denial of Service. Consequently, "confidentiality"—prevention of unauthorized disclosure of information to unauthorized individuals; "integrity"—that information has not been modified except by the authorized individuals; and "availability"—that the stored information remains available and the service is not denied, are the cardinal features of cyber security measures [13]. The mechanisms organizations use to protect, identify, and address the issues of cyber security incidents are described below.

4.6.3 Measures for Ensuring Cyber Security

Organizations and individuals use a host of measures for ensuring cyber security, including data labeling, security controls, and security testing [28].

4.6.3.1 Data Labeling

Organizations label their data, as per the importance, as "confidential" or "public," which determines the type of protection required for the data and the data storage, sharing, and access mechanisms. Consequently, a laptop with important data and a banking web network responsible for transmitting credit card transactions can both be labeled as Confidential. By contrast, an organization's web portal for the public would be classified as Public, thus requiring different levels of cyber security for the online users of internet.

4.6.3.2 Security Controls

Organizations and individuals use the classification of data to determine the required security controls, which may be (1) preventative, (2) detective, and (3) corrective [29]. Preventive controls aim at prevention of occurrence of any incident and include firewalls and door locks. Detective controls aim at the identification of an ongoing incidence and include intrusion detective devices and smoke detectors. Corrective controls are applied to deal with the damage done by an incident so that the organization quickly returns to normal working. This includes fire extinguishers and incident response plans.

4.6.3.3 Security Testing

Organizations use the means of security testing to ensure that the security control mechanisms put in place are effective in the wake of any cyber-attack. The commonly used security testing activities are social engineering and penetration testing [30]. Increasingly, the government and private sector organizations in many countries are evaluating their security response on a periodic basis and are thus certified by the cyber security professionals. Cyber security is also ensured through compliance of cyber security laws, which are discussed below.

4.6.4 Cyber Security Laws

Governments have passed cyber security laws. In 1996, the US government passed the Health Insurance Portability and Accountability Act (HIPAA). This was the first time cyber security became the requirement for healthcare organizations in the United States. Since then, countless new laws have been passed. Laws are meant to ensure appropriate handling of information and data in the cyberspace. Noncompliance to these regulations enhances the risks of hacking and data theft of organizations and individuals, thus necessitating proportionate penalties from the regulators [15,31].

4.6.5 Main Features of the Pakistan's Prevention of Electronic Crimes Act 2016

The Prevention of Electronic Crimes Act came into force on August 19, 2016. Chapter two of the Act gives a list of cyber security offenses and punishments that includes:

Unauthorized access to information system or data,
Copying and transmission of data,
Interference with information systems or data,
Access to critical infrastructure information system or copying and transmission of data,
Glorification of offense,
Cyber terrorism,
Hate speech,
Recruitment, funding, and planning of terrorism,
Electronic frauds, forgery, theft,
Tempering of communication equipment,
Unauthorized interception,
Offenses against dignity and modesty of a person,
Child pornography,
Cyber stalking,
Spamming and spoofing.

The punishments range from three months and 50,000 rupees for an offense of unauthorized access to information systems or data, to seven years of imprisonment for hate speech, to 14 years of imprisonment or fine up to 50 million rupees, or both for cyber terrorism, including coercing, intimidation, or creation of "sense of fear, panic, or insecurity in Government or public or a section of the public or community or sect or create a sense of fear or insecurity in society" or even further interfaith, sectarian hatred or promote the objectives of proscribed organizations.

The Act has been criticized on the grounds of impinging on human rights and freedom. However, with less than a year from its inception, it is far from being fully implemented and no attempt has been made to raise awareness, to make people aware of the threats of cyber security and knowledge of the available avenues and laws to tackle these threats.

4.7 Cyber Security—Some Recent Incidents

This section provides a snippet of some recent cyber security incidents in the world [32]. These incidents point to the challenges of securing cyber security to which Pakistan is not immune.

4.7.1 Cyber-Attacks for Cyber Ransom: Ransomware

Hacking for the sake of ransom is becoming a lucrative cybercrime. There has been 172% increase in the number of ransomware attackers in the first half of 2016 alone. Healthcare and education institutions have repeatedly been targeted. For example, in February 2016, there was a cyber-attack on the Hollywood Presbyterian Medical Center, which for a week disrupted its routine daily operations, documentations, and medical scans. The hospital paid 40 Bitcoins (17,000$) to restore the systems. In March 2016, in the United States, Med Star, a network of ten hospitals with 250 outpatient centers, was paralyzed and 45 Bitcoins (19,000$) of ransom was demanded. In May 2016, University of Calgary (Canada) paid 20,000 dollars as ransom to avoid exposure of its critical data and research. In another incident, on Thanksgiving weekend, the ticketing system of San Francisco Municipal Transportation Agency (SFMTA) was attacked to compromise 2,000 computer systems and 30 gigabytes of data was stolen. The organization had to take the ticketing system offline and the commuters used the trains for free, resulting in the loss of revenue.

4.7.2 Cyber-Attacks on Bangladesh Bank & Leoni

At times, hackers use legitimate email accounts to trick financial personnel of the multinational companies to wire transfer the funds to their accounts. This is referred to as Business Email Compromise (BEC). Since 2015, there has been 270% increase in the number of victims of BEC who lost substantial amounts. Another variation of BEC is Business Process Compromise (BPC) in which rather than compromising the emails, business processes, practices, and organizational loopholes are exploited for financial gains. Attackers first gain foothold in the target organization by finding a point of compromise and then moving laterally in the organization to explore vulnerabilities. For example, in February 2016, in a unique cyber-attack, hackers targeted Bangladesh Bank to inflict a whopping loss of $81m. They were able to transfer these funds to accounts in the Philippines and Sri Lanka by first gaining access to the Bank and then generating requests to the Federal Reserve Bank of New York to transfer funds to designated accounts. They avoided detection by synchronizing such requests with the weekends and also by tempering with bank printing system so that SWIFT receipts of funds transfers are not printed. The SWIFT codes were also gained access too by the hackers in 2015 to attack the Ecuadorean bank Banco del Austro who lost $12m. The money was transferred to accounts in New York, Los Angeles, Dubai, and Hong Kong. Later, SWIFT asked its clients to update their software to thwart any cyber-attack. In another incident, in August 2016, the Chief Financial Officer of Leoni—a cable manufacturing company in Germany, who was based in Romania, acted on a spoofed email that appeared to be sent by the top executive, and also seemed to comply with company's policy, tricked into transferring a staggering amount of €40m.

4.7.3 Distributed Denial of Service (DDoS) Attacks: Mirai

DDoS attacks have commonly been conducted through a malware Mirai, which has affected CCTV systems, home networking systems, and the routers. It turns Internet of Things (IoT) devices into bots [33]. When a cyber security blogger wrote about DDoS attacks, he had to face 620 Gbps of malicious traffic, and as a result, his site was shut down for many days. However, after help from Google, he was able to go back online. Later on, DDoS attack on Dyn, a DNS provider, affected almost 100,000 IoT devices, Twitter, Reddit, Netflix, and Spotify, as well as millions of users on the US East Coast. In November, five Russian banks encountered DDoS attacks for two days involving 24,000 computers in 30 countries.

4.7.4 Cyber-Attack on Ukrainian Power Grid

In a first cyber malware attack on any industrial facility, Ukrainian power grid was targeted. The attackers through hijacked Virtual Private Networks (VPNs) accessed Supervisory Control and Data Acquisition (SCADA) to control the electric grid. The attack that caused unscheduled power outages lasted for three hours; effected three power distribution companies and 250,000 customers; and rendered the operators powerless. Later, it was identified that the attackers had used Black Energy malware and the Kill Disk module in it to wipe some system, thus making others inoperable.

4.7.5 Yahoo (2013) Cyber-Attack

On December 14, 2016, in an acknowledgment of a hitherto unprecedented theft of the personal data of the online users of internet throughout the world, including users from Pakistan, Yahoo disclosed that the investigations by the law enforcement had revealed that in August 2013, it has been a target of successful malicious cyber-attack by the state-sponsored actors, which compromised data of more than 1 billion accounts [34,35]. As a result, data such as email addresses, user names, passwords, mobile phone numbers, and security questions and answers were stolen. It is believed that financial data including bank account details and the credit card numbers were not stolen, but the stolen encrypted passwords were likely to be compromised. Yahoo notified the affected users and provided users some basic information regarding secure use of Yahoo mail [34].

This incident shows the vulnerability of email accounts, data transmitted through emails, the extent to which individuals and state-sponsored actors could go to illegally gain access to the data of online users of internet, and the need for awareness of cyber security.

4.7.6 Democratic National Committee (DNC) under Attack

Hackers targeted the DNC of the Democratic Party and stole 19,000 emails from the server. The attacks were claimed by Romanian hacker group Guccifer 2.0. Debbie Wasserman Schultz, the chair of the DNC, resigned. There have been speculations that the leaks were aimed at influencing the outcome of US Presidential elections in favor of Mr. Trump and against Hillary Clinton.

4.7.7 Smart Phones Hacking: Apple Zero-Days

Smart phones have also become the target of hacking in 2016. Trident, a mobile phone malware, was accidentally discovered when in UAE someone found suspicious text messages and got it investigated from Citizen Lab Toronto and Lookout, a mobile security firm [36]. Citizen Lab and Lookout found that the "Trident" is based on the flaws in Apple's iOS, which is used by an attacker to silently jailbreak into the phone device and get access to communication data from Gmail, Facebook, Skype, WhatsApp, etc. Lookout assisted Apple to fix it and launch an iOS update, 9.3.5. Apple asked its customers to update the latest version of the iOS. It is believed that persons with sensitive information, government officials, businessmen, and other persons of interest are potential future victims. The threat is more likely if the individuals are running malicious software, noncompliant applications, and compromised operating systems, hence public sector organizations; government agencies and policy makers need to be aware of cyber security threats.

4.7.8 Fake Facebook Accounts

In 2012, Facebook reported that it had an estimated 83.09 million fake accounts [37]. Among these, 45.8 million (4.8%) were considered duplicate accounts; 22.9 million (2.4%) are misclassified accounts, non-human, or company accounts which should actually be pages; and 14.3 million (1.5%) of all active accounts are maliciously created for spamming. Facebook believes that false accounts are more in the developing countries than in developed countries.

4.7.9 Cyber Security Incidents in Pakistan

Pakistan has also faced many cyber security attacks. Most of the attacks were launched by the Indian hackers, but new reports have surfaced suggesting that governments such as that of the United States had also got access to the data of Pakistani mobile service provider. Few of the cyber-attacks are as follows: on October 07, 2016, a website reported that Indian hackers allegedly, a part of Telangana cyber warrior, claimed to have locked hundreds of Pakistani government computers and refused to accept Bitcoins to unlock the systems to show their patriotism [38]. According to the same website, the Pakistan Haxors Crew, a group of Pakistani hackers, had claimed to have defaced 7,051 Indian websites as a retaliation to the alleged Indian surgical strike in Pakistani Kashmir. On February 14, 2017, Indian hackers, IND 3MB3R, hacked the website of Lahore district government showing images of a sadhu and an insignia of the Indian Army (Lahore District Govt 2017). Wikileaks reported that according to Twitter account "x0rz," the US National Security Agency (NSA) had got access to Pakistan's Mobilink's GSM network. Mobilink's Chief Technical Officer, Khalid Shehzad, confirmed that it had happened in 2006 with no known implications [39]. On April 25, 2017, it was reported that Indian hackers hacked 30 Pakistani websites to protest the death sentence awarded to Indian spy Kulbhushan Jadhav who was arrested by the Pakistani authorities on charges of sabotage and espionage in Balochistan and Karachi [40]. The report also claimed that Telangana Cyber Warrior, another hackers group, claimed infecting a government hospital in Karachi and stating that "We have their entire computer network under our control. We can modify patients' records and monitor their health condition." Kerala Cyber Warriors hacked the portal of Pakistan Academy for Rural Development, showing a message "justice for Kulbhushan Jadhav Ji." On April 26, 2017, Indian hackers, called Ind_C0d3r and Mr Sh3ll, hacked the website of Pakistan People's Party, which started showing Indian Flag and an inflammatory message "Stop barking for Kashmir" [41].

In this context, a survey was designed and conducted to gauge the awareness regarding cyber security among senior public sector executives in Pakistan. The following section presents the key findings from the survey conducted in June 2017.

4.8 Results of the Survey on Awareness of Cyber Security

The survey was given to a total of 55 senior public servants to gauge their awareness and the organizational preparedness about the threats of the cyber security, and 46 of them responded—a response rate of 83.6%. The responses are organized in the following section, according to the key aspects explored in the questionnaire.

4.8.1 Participants Receive Spam Emails

An alarming 56.5% of participants (26) reported to have received emails that appeared to be from their contacts, but were not from their contacts but spam emails. The remaining 43.5% of the participants (20) reported not to have received any such emails. A vast majority (35), that is, 76%, noted to have received spam emails apparently from their friends, informing them that she/he was stuck in Europe/United States and needed financial help from them urgently, while remaining 24% participants (11) have not received such emails.

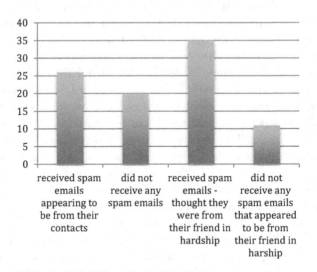

4.8.2 Participants Use Private Emails

Many public servants have noted that they use personal emails for official communication. Although some ministries have the provision for office emails, it is rarely mandated to use office emails; hence, external private email accounts are being used by all the participants of the study. 54% of the participants (25) reported to be using Gmail account (2 among them also use Hotmail), 28.3% (13) use Yahoo email (three among these also use Gmail), and 17.4% (8) use Hotmail account. Altogether, 100% respondents use an email server hosted in other countries.

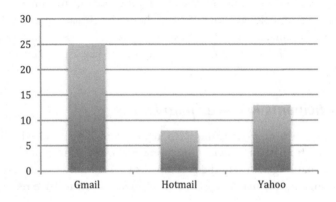

4.8.3 Insufficient Knowledge of Securing Personal Emails

A total of 63% of the participants (29) reported of not knowing the different methods of securing their email accounts. Only 37% (17) self-reported to know such methods. Majority of the participants, that is, 80.4% (37), reported that they properly sign out after using their email account; only 19.6% (9) reported not to be properly signing out. An overwhelming majority, that is, 63% (29), reported to have logged into their email accounts using computers from a library, cyber café, or hotel lobby; only 37% (17) reported to have never done so.

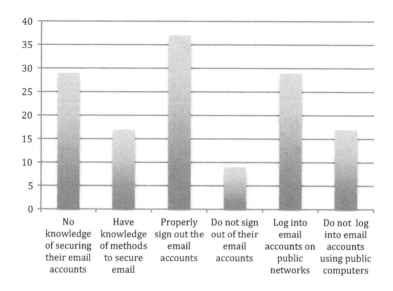

4.8.4 Use of Weak and Same Passwords

Majority of the participants, that is, 67.4% (31), reported to be using the strong password such as Dp0si#Z$2 made up of uppercase, lowercase, and numerals, but 32.6% (15) participants reported to be using the weak passwords such as My Secret and Abc123 (10). Majority of the participants, that is, 69.5% (32), reported that they do not share their email account/password with family members/staff; however, 30.5% (14) participants reported to have shared their email account/password with family members/staff. Whopping, that is, 60.9% (28), participants reported that they find it easy to make one password for many different accounts, including Facebook and email account, because it is easy to remember. Only 39.1% (18) participants reported not to have done so. Majority of the participants, 69.5% (32), consider a combination of upper and lowercase letters mixed with numbers and symbols, a good way to create password. 10.9% (5) participants consider using children's names, 6.5% (3) participants consider using look-alike substitutions of numbers or symbols, and 8.7% (4) consider using common names or words from the dictionary.

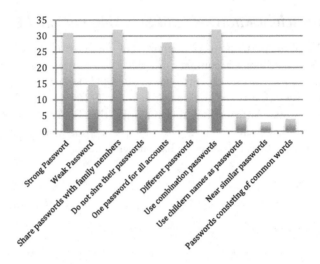

4.8.5 Permanently Deleted Files Are Recoverable

Majority of the participants, 71.7% (33), do not believe that deleting files from laptop or even formatting hard drive all the information on it is permanently lost. Only 19.5% (9) participants believe deleting files permanently deletes files from laptop. 8.7% (4) participants did not give any answer.

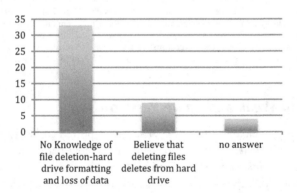

4.8.6 Role of Antivirus

Majority of the participants, 65.2% (30), do not believe that installing antivirus on a computer/laptop will stop all viruses, worms, Trojans that infect the computer/laptop. 28.3% (13) believed that installing antivirus will stop and 6.5% (3) gave no response. Majority of the participants, that is, 67.4% (31), reported that antivirus was currently installed, updated, and enabled on their computer/laptop. 17.4% (8) reported that it was not installed. 8.7% (4) were not sure whether antivirus was installed or not. And 6.5% (3) did not know what antivirus was.

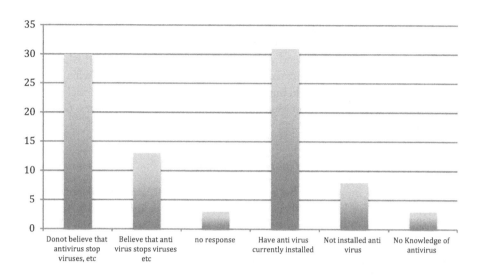

4.8.7 Lack of Institutional Policies to Access Internet at Workplace

Half of the participants, 50% (23), reported that there were no policies on which websites they could visit during work and they could visit whatever websites they wanted to while at work. 28.3% thought that perhaps there were policies regarding what websites could or could not be accessed at work but they were not sure what those policies were. 17.4% were confident that policies existed and that they knew and understood those policies. 4.3% (4) did not respond to the question.

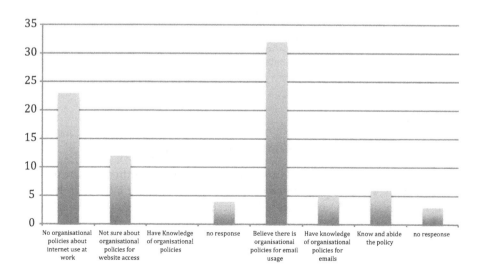

An overwhelming majority of the participants, that is, 69.5% (32), reported that in their department/organization, there are no policies as to how they can and cannot use email. They could send whatever emails they wanted to and whomever they wanted to. 10.8% (5) reported that there are policies limiting what emails could (not) be sent while at work, but they have no knowledge of such policies. 13% (6) reported that such policies existed and they understood these policies. 6.5% (3) participants gave no answer.

4.8.8 Little Support in Case of Hacking

An overwhelming majority of the participants, 63% (29), reported of not knowing what to look for to see if their computer is hacked or infected. 58.7 percent of the participants (27) reported that they did not know whom to contact in case their email account is hacked or if their computer is infected. 6.5% (3) participants reported that there was no cyber security team in their organization/department. 4.3% (2) participants chose not to respond. This suggests that the participants are more likely to use a compromised device, which can expose the organization to security breach and compromise.

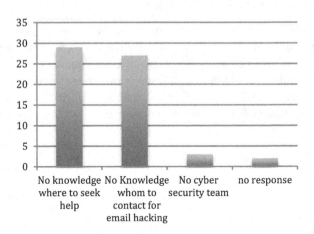

4.9 Discussion

4.9.1 Email Spoofing Online Users of Internet

A key finding of the research has been that the cyber security threat to the public sector executives is real. As pointed out in the results section, more than half of the participants (56.5%) reported to have received spam emails that prima facie were from their contacts but were in fact spams. Similarly, an overwhelming majority of the participants (76%) had received emails apparently from their friends informing them that she/he was stuck in Europe/United States and needed financial help from them. These emails contain a link to some website, and a click on that link may unintentionally download malicious malware on the computer. Hackers also use contact lists from hacked emails to send further emails, hence creating a chain and thus compromising data in the email of the contact. In cyber security parlance, it is called email spoofing. Spammers do it in spam campaigns because people would open email thinking it is from a legitimate contact.

4.9.2 Private Emails Pose Risks to Confidentiality

A key finding of the study is that in the absence of official email accounts, participants use private email accounts for the official business at the National Management College. The study found that all participants used an email server hosted by other countries. Gmail was used by 54% of the participants (2 of them also use Hotmail), Yahoo was used by 28.3% of the participants (three of them also use Gmail), and Hotmail was used by 17.4% of the participants. These emails are connected to the servers, which are based in other countries. Thus, every time a participant using a Yahoo, Gmail, or a Hotmail account presses the send button to send a study report or a synopsis, she/he is unconsciously pushing information out of the physical boundaries of Pakistan. Given the amount of data and information generated in various ministries and departments, this is the most serious security risk identified by the study. It is not difficult to imagine the quantum of the problem if public policy makers in all the government ministries and departments are using generic email accounts.

4.9.3 Little Knowledge of the Risks of Public Computers

Although majority of the participants (80.4%) reported to be properly signing out their email, 63% of the participants said they had no knowledge of how to secure their email accounts. An overwhelming majority (63%) reported to have logged into their email accounts using computers from a library, cyber café, or hotel lobby. This exposes them to serious cyber security threats, as there is little or no knowledge of the users of the purpose for which public computer has previously been used for. There is high probability that such computers may be infected with spyware and virus, hence a security threat to official communication and data.

4.9.4 Weak/Same Passwords Mean Weak Security

The study found that around 32.6% of the participants reported to be using the weak passwords, which puts them at serious risk. Similarly, 30.5% of the participants reported to have shared their email account/password with family members/staff, which is the serious cause of concern. Interestingly, a significant percentage of participants (60.9%) reported of using one password for many different accounts, including Facebook and email account, due to the ease of remembering it. Similarly, 10.9% participants use children's names, 6.5% participants use look-alike substitutions of numbers or symbols, and 8.7% use common names or words from the dictionary. Now these practices make participants vulnerable to serious cyber security threats. For hackers, it is not difficult to find the personal information, which is used as password [14,42]. It is a common knowledge that personal information such as first name, surname, date of birth, and even mobile phone numbers is required for making an account with Facebook, Twitter, and LinkedIn. The same information is used for opening a bank account or registering with a health service or practitioner. Hackers use this information for cracking the account, and the one who uses this information is more vulnerable to be hacked. This study suggests that unfortunately, public sector executives of Pakistan potentially fall in this category of being vulnerable to such threats.

4.9.5 Emptying Bin Does Not Delete Files

The study found that 19.5% of the participants believed that once a file is deleted from a computer/laptop, it gets permanently deleted, and the remaining (8.7%) participants did not respond, which is equally concerning. A deleted file is still stored on the hard drive of the computer even when the Recycle Bin is emptied. Therefore, it is very easy for the hackers or anyone who can access the computer to recover the files, which an individual believes to have been permanently deleted. This may compromise the confidentiality of data in situations when the same computer is used by other people or when the computer changes hands in public offices or is replaced.

4.9.6 The Limits of Antivirus & Technical Solutions

Almost 28.3% reported that installing antivirus on computer/laptop will prevent all viruses, worms, and Trojans, from infecting the computer/laptop. 17.4% reported that antivirus was not currently installed, updated, and enabled on their computer/laptop, 8.7% were not sure whether antivirus was installed or not, and 6.5% had no idea of what an antivirus is. This suggests the need for creating awareness about antivirus software. They are also unaware of the fact that antivirus software only provides protection from the known viruses and provides no protection against new ones, hence the need to continually upgrade the software. No matter how good a technical solution is, security breach requires only a weak link. Hackers often try to trick the users to click a malicious link on a website or open an infected attachment. Thus, the user itself is tricked into bypassing the firewalls and antivirus leading to compromised security of the computer. Therefore, along with the technical solutions, users require tailored security awareness training [10,21,43].

4.9.7 Lack of Policies to Access the Internet at Workplaces

A key finding of the study is that half of the participants (50%) reported that there were no policies on which websites they could visit during work, and hence, they could visit whatever websites they wanted to while at work. Similarly, many thought (28.3%) that perhaps policies regarding what websites could or could not be accessed at work were there but they were not sure what those policies were. Also, a significant majority of the participants (69.5%) reported that since there were no policies in their department/organization as to how they can and cannot use email, they can send emails from workplace to anyone. Also, 10.8% reported that there are policies limiting what emails could (not) be sent while at work, but they have no knowledge of such policies. Now this lack of policies and not being sure about the existence of policies pose serious risks to both the individuals and the organizations they are employed in. This is a serious lack of oversight by the organizations.

Organizations and businesses adopt Internet Usage Policy especially in the private sector, similar to the policies implemented in the public sector in many developing countries like UK, and Australia. The common features of such policies require that the employees use internet for the business/work purposes only. The misuse of internet necessitates disciplinary action against the individuals. Emails are to be in line with the organizational policies and must not contain any defamatory and offensive content. The emails and all communications are considered the property of the organizations, with rights of access reserved. Employees are advised not to use the email for personal messages. The emails generated from the organizational account often contain the expressed consent that the organization reserves the right to review and monitor the internet use by the employees.

4.9.8 Lack of Knowledge about Hacking & Institutional Support

Another important finding of the study is that a significant majority of the participants (63%) had no knowledge of the signs to look for to see if their computer is hacked or infected. This suggests that the participants who had no idea of the potential signs to look for run the risk of continuing to use a security-compromised and affected device, thus seriously exposing the individuals and their organizations to security breaches, compromising the official information and data. It becomes even more serious due to the finding that in the study, an overwhelming majority (58.7%) of the participants reportedly have no knowledge of the relevant policy or office that can be contacted if their email account is hacked or office computer is infected. These participants pose a significant risk to the organization for they can potentially continue to use the security-compromised device, which may result in a further breach and compromise the security of official information as well as data.

4.10 Conclusions

The results of the case study survey and the discussion suggest that the organizations to which the participants of the survey belong as well as these public sector executives themselves are at high risk of cyber security attacks. The significant majority of these public sector executives are neither aware of cyber security threats nor do they practice the security standards (if there are any). In fact, organizations have not been found to have cyber security policies, which may guide the attitude and behavior of the public sector executives of Pakistan. Therefore, due to a host of factors, the public sector executives in Pakistan unwittingly engage in activities and practices, which puts them at the risk of enhanced cyber-attacks. Based on the findings of this case study, one can argue for similar research in developing countries, to understand cyber security challenges faced by senior public service personnel.

4.11 Recommendations

In view of the results and discussion, the following recommendations are presented.

4.11.1 Awareness of Spam Emails and Spoofing

In view of the high probability of malicious email attacks, it is important that the public sector executives are made aware of the dangers posed by the spam emails and malicious websites. Public service may consider engaging National Information Technology Board (NITB) as well as provincial information technologies boards to provide training.

4.11.2 Mandatory Use of Official Emails by Participants

Since this study unequivocally sensitizes about the threat posed by the use of private emails, it is important that steps may be taken so that the public sector executives only use official email account. This may be achieved in consultation with the NITB and the Ministry of Information Technology. The latter must ensure that all departments and ministries are using the official email account with officially authorized government of Pakistan server.

4.11.3 Awareness Training for Participants

It is recommended that all ministries and relevant departments organize and mandate awareness training, which includes the issues raised in this study including the dangers inherent in the use of public computers, weak passwords including the dangers associated with the use of personal information in passwords, and more generally awareness about the limitations of antivirus and deletion of files. It is stressed that such training sessions will help combat cyber security threats. More generally, such awareness can also be created through the inclusion of such topics in the training programs for public sector executives.

4.11.4 Policies for Use of the Internet at Workplaces

The government needs to formulate the policies on the use of internet at workplace. Such policies may monitor and restrict the use of internet at workplace to the official business as much as these make the emails comply with the organizational mandate and policies. Similarly, such policies also need to create mechanisms of support for the individuals in need or at risk of cyber security breaches.

Bibliography

1. "ISO/IEC 27032:2012 Information Technology - Security Techniques - Guidelines for Cybersecurity", accessed 20 April 2017, www.iso27001security.com/html/27032. html.
2. H. O. Hundley, and R. H. Anderson. "Emerging Challenge: Security and Safety in Cyberspace," IEEE Technology and Society. *IEEE Technology and Society Magazine*, Vol. 14, Issue 4, pp. 19–28, 2002.
3. Paul N. Edwards, *The Closed World: Computers and the Politics of Discourse in Cold War America*. Cambridge, MA: MIT Press, 1996.
4. Alastair MacGibbon. "Cyber Security: Threats and Responses in the Information Age," Australian Strategic Policy Institute, Special Report - Issue 26, Barton, 2009.
5. Lene Hansen and Helen Nissenbaum. "Digital Disaster, Cyber Security, and the Copenhagen School," *International Studies Quarterly*, Vol. 53, pp. 1155–1175, 2009.
6. Won Kim, Ok-Ran Jeong, Chulyun Kim, and Jungmin So. "The Dark Side of the Internet: Attacks, Costs and Responses," *Information Systems*, Vol. 36, pp. 675–705, 2011.
7. Ross Gore, Jose Padilla, and Saikou Diallo. "Markov Chain Modeling of Cyber Threats," *The Journal of Defense Modeling and Simulation*, Vol. 14, no. 3, pp. 233–244, 2017.
8. Joseph S. Nye Jr. *The Future of Power*. New York: Public Affairs, P. 123, 2011.
9. Philip Elliott and Zeke J. Miller. "Alleged Russian Hack of Democrats Reshapes Presidential Election Fight," *Time*, 26 July 2016. http://time.com/4422379/russian-hack-democratic-convention-putin/, accessed 12 April 2018.
10. Michael E. Whiteman. "Security Policy: From Design to Maintenance," in Detmar W. Straub, Seymour Goodman, and Richard L. Baskerville (eds.), *Information Security: Policy, Processes, and Practices*. New York: M. E. Sharpe, pp. 127–151, 2008.
11. "National Cyber Security Strategy 2016–2021," 2016. www.gov.uk/government/uploads/system/uploads/attachment_data/file/567242/national_cyber_security_strategy_2016.pdf, accessed 12 April 2018.
12. "The Prevention of Electronic Crimes Act," 2016, Act No. XL of 2016, Government of Pakistan. www.na.gov.pk/uploads/documents/1472635250_246.pdf, accessed 12 April 2018.

13. Deborah L. Wheeler. "Understanding Cyber Threat," in Kim Andreasson (ed.), *Cyber Security: Public Sector Threats and Responses*. New York: CRC Press, pp. 27–53, 2011.

14. S. M. Furnell, P. Bryant, and Andrew D. Phippen. "Assessing the Security Perceptions of Personal Internet Users," *Computers & Security*, Vol. 26, No. 5, pp. 410–417, 2007.

15. "2015 Cyber Security Survey: Major Australian Businesses," Canberra: Australian Cyber Security Centre.

16. Monica Whitty, James Doodson, Sadie Creese, and Duncan Hodges. "Individual Differences in Cyber Security Behaviors: An Examination of Who is Sharing Passwords," *Cyberpsychology, Behavior, and Social Networking*, Vol. 18, No. 1, pp. 3–7, 2015.

17. Nur Azha Putra and Kevin Punzalan. "Cyber Security," in Mely Caballero-Anthony and Alistair D. B. Cook (eds.), *Non-Traditional Security in Asia: Issues, Challenges and Framework for Action*. Singapore: Institute of Southeast Asian Studies, pp. 267–289, 2013.

18. Hilarie Orman. "The Morris Worm: A Fifteen-Year Perspective," *IEEE Security & Privacy*, Vol. 99, No. 5, pp. 35–43, 2003.

19. Mark T. Maybury. "Toward Principles of Cyberspace Security," in Jean-Loup Richet (ed.), *Cybersecurity Policies and Strategies for Cyberwarfare Prevention*. Hershey, PA: Information Science Reference, an imprint of IGI Global, pp. 1–12, 2015.

20. Karsten Friis and Jens Ringsmose (eds.), *Conflict in Cyber Space: Theoretical, Strategic and Legal Perspectives*. New York: Routledge, 2016.

21. Michael D. Coovert, Rachel Deibelbis, and Randy Borun. "Factors Influencing the Human-Technology Interface for Effective Cyber Security Performance," in Stephen J. Zaccaro, Reeshad S. Dalal, Lois E. Tetrick, and Julie A. Steinke (eds.), *Psychosocial Dynamics of Cyber Security*. New York: Routledge, pp. 267–290, 2016.

22. Shima D. Keene. *Threat Finance: Disconnecting the Lifeline of Organised Crime and Terrorism*. Surrey: Gower, 2016.

23. Stephen J. Zaccaro, Amber K. Hargrove, Tiffani R. Chen, Kristin M. Repchick, Tracy McCausland. "A Comprehensive Multilevel Taxonomy of Cyber Security incident report Performance," in Stephen J. Zaccaro, Reeshad S. Dalal, Lois E. Tetrick, and Julie A. Steinke (eds.), *Psychosocial Dynamics of Cyber Security*. New York: Routledge, pp. 13–55, 2016.

24. Jeffrey Carr. *Inside Cyber Warfare: Mapping the Cyber Underworld*. California: O'Reilly Media, 2011.

25. Noah Rayman. "The World's Top 5 Cybercrime Hotspots," 2014, http://time.com/3087768/the-worlds-5-cybercrime-hotspots/, accessed on 10 May 2018.

26. Christian W. Probst and Jeffrey Hunker. "Insiders and Insider Threats An Overview of Definitions and Mitigation Techniques," *Journal of Wireless Mobile Networks, Ubiquitous Computing, and Dependable Applications*, Vol. 2, No. 1, pp. 4–27, 2010, http://isyou.info/jowua/papers/jowua-v2n1-1.pdf, accessed 10 May 2018.

27. Will Gragido and John Pirc. *Cybercrime and Espionage: An Analysis of Subversive Multi-Vector Threats*. Boston: Elsevier, 2011.

28. Mark S. Merkow and Jim Breithaupt. *Computer Security Assurance Using the Common Criteria*. New York: Thomson, 2004.

29. Robert M. Clark and Simon Hakim (eds.), *Cyber-Physical Security: Protecting Critical Infrastructure at the State and Local Level*. New York: Springer, 2016.

30. Rebecca Herold. *Managing an Information Security and Privacy Awareness and Training Program*. London: CRC Press, 2010.

31. Elaine C. Kamarck. "The Cyber Security Policy Challenge: The Tyranny of Geography," in Kim Andreasson (ed.), *Cyber Security: Public Sector Threats and Responses*. New York: CRC Press, pp. 109–125, 2011.

32. "A Rundown of the Biggest Cybersecurity Incidents of 2016," 18 December 2016, www.trendmicro.com/vinfo/au/security/news/cyber-attacks/a-rundown-of-the-biggest-cybersecurity-incidents-of-2016.

33. George Lucas. *Ethics and Cyber Warfare: The Quest for Responsible Security in the Age of Digital Warfare*. New York: Oxford University Press, 2017.

34. Bob Lord. "Important Security Information for Yahoo Users," 2016, https://yahoo.tumblr.com/post/154479236569/important-security-information-for-yahoo-users.

35. "Yahoo Discloses 2013 Breach that Exposed Over One Billion Accounts," www.trendmicro.com/vinfo/us/security/news/cyber-attacks/yahoo-discloses-2013-breach-exposed-over-1billion-accounts, accessed 15 April 2018.

36. "A Rundown of the Biggest Cybersecurity Incidents of 2016," www.trendmicro.com/vinfo/au/security/news/cyber-attacks/a-rundown-of-the-biggest-cybersecurity-incidents-of-2016, accessed 15 April 2018.

37. Heather Kelly. "83 Million Facebook Accounts are Fakes and Dupes," 2012, http://edition.cnn.com/2012/08/02/tech/social-media/facebook-fake-accounts/, accessed 15 April 2018.

38. Shashnak Shekhar. "Patriotic Indian Hackers Lock Pakistani Websites and Refuse to Give Back the Key," *The Daily Mail*, 7 Oct 2016, www.dailymail.co.uk/indiahome/indianews/article-3825751/Patriotic-Indian-hackers-lock-Pakistani-websites-refuse-key.html, accessed 15 April 2018.

39. "NSA Hacked Pakistani Mobile System: Wikileaks," *The Dawn*, 10 April 2017, www.dawn.com/news/1326104, accessed 15 April 2018.

40. "Indian Hackers Target 30 Pakistani Sites to Avenge Jadhav's Death Sentence," *The Nation*, 25 April 2017, http://nation.com.pk/national/25-Apr-2017/indian-hackers-target-30-pakistani-sites-to-avenge-jadhav-s-death-sentence, accessed 15 April 2018.

41. "PPP Website defaced by Indian hackers," *The Express Tribune*, 26 April 2017 from https://tribune.com.pk/story/1393704/ppp-website-defaced-indian-hackers/, accessed 15 April 2018.

42. K. Y. Williams, Dana-Marie Thomas, and Latoya N. Johnson. "The Value of Personal Information," in Eugenie de Silva (ed.), *National Security and Counterintelligence in the Era of Cyber Espionage*. Hershey, PA: Information Science Reference, an imprint of IGI Global, pp. 161–180, 2016.

43. Michael Robinson, Kevin Jones, and Helge Janicke. "Cyber Warfare: The State of the Art," in Jean-Loup Richet (ed.), *Cybersecurity Policies and Strategies for Cyberwarfare Prevention*. Hershey, PA: Information Science Reference, an imprint of IGI Global, pp. 13–36, 2015.

Chapter 5

A Benefits Realization Approach to Cyber Security Projects Implementation in Public Sector Organizations

Munir A. Saeed
UNSW Canberra

Tahmina Rashid
University of Canberra

Mohiuddin Ahmed
Edith Cowan University

Contents

5.1 Introduction

WikiLeaks publishing nearly 10 million documents in a decade (2006–2016), hacking into the former US Secretary of State and US presidential candidate Hillary Clinton's emails, rumors of Russian interference in the US elections in 2017, news reports of hacking into computing system of Australian Parliament House, and stealing of the design of yet under construction submarines for Australia have created panic in various capitals and governments are allocating hefty budgets for consolidating defenses against cyber breaches. However, the cyber threats are real and will increase as the developed world is fast moving towards the Internet of Everything (IoE) environment. According to the findings of Telstra Security Report 2019, 65% of Australian businesses suffered disruptions due to cyber breaches last year and 89% said they had breaches which had gone unnoticed. According to the Australian Cyber Security Centre Report (2017), foreign adversary states have the greatest ability to threaten Australian networks and over a period of one-year extensive activities have been detected against the Australian government and the private sector networks in support of their national security objectives, foreign policy, and economic interests. According to this report, more than half of these attacks were directed at the industry and 28% attacks targeted Commonwealth and States government agencies. Australian private sector is a popular target of cyber espionage attacks, seeking valuable information such as intellectual property, latest research, and new technologies, which subsequently adversely affects the competitive advantage of Australian companies. ACSC Report (2017) further states that foreign investment in the Australian private sector is one of the key motivators for the adversaries to launch cyber espionage attacks against Australian business interests. According to the Guardian newspaper (2019), China is a suspect behind breaches into Australian National University (ANU), in which students and staff data such as bank accounts, passport, tax, and academic details of last 19 years have been stolen. Landis-Handley (2019) states that data stolen at ANU belongs to over 200,000 students and staff. Quoting intelligence officials, the Guardian newspaper writes, the stolen details can be used to recruit the current and former students as informants. Similarly, according to the Australian Broadcasting Corporation (ABC), Perth-based Naval shipbuilder Austal and Australian Parliament computer network were attacked by sophisticated state actors. Therefore, over 80% of the organizations expend about 20% of their Information Technology (IT) budget on cybersecurity. With the cyber threats ever increasing, the private and the public sectors will implement new cyber technologies through projects. However, the rate of success in IT projects has been low, and cybersecurity projects are expected to suffer from challenges.

There is an urgent need to apply a benefits realization (BR) approach to all projects for cybersecurity. Current project management (PM) approaches are output oriented but BR approach is outcomes based. Since the 1980s as the PM matured as a discipline, it defined project success on the successful management of scope, time, and cost. But over the years, PM professionals and academics have realized that output-based PM approach is flawed, as it is only delivery based. Therefore, since a decade and a half, a new approach to PM has emerged, it is known as benefits realization, which focuses on the outcomes realized from the project outputs. In this chapter, the authors are proposing a similar approach to cybersecurity projects so that investments made into fortifying defenses against cyber-attacks result into real and realizable benefits to the governments and private businesses.

5.2 Project Management

Projects have been employed as a tool to achieve strategic objectives by various industries since the 1950s starting with the defense industry in the United States. The important role played by projects as an important tool can be gauged from the share of projects in the global economic activity. According to Bredillet et al. (2013), at the global level, more than 20% of the economic activity takes place in the shape of projects, and in the fast developing economies, the share of economic activity constitutes more than 30%. Similarly, nearly an 11% global GDP totaling $48 trillion is produced through projects, which indicates the contribution of projects to the value creation in the world (Bredillet et al. 2013).

The Project Management Institute (PMI) defines a project as a temporary endeavor to produce a unique service, a product, or result. Oisen (1971) writes, PM is the application of a number of tools and techniques, to manage the use of diverse resources in order to perform a task within the constraints of time, cost, and quality. The IPMA Competence Baseline (I. C. B. ND) looks at project as a cost and time-constrained operation, initiated to achieve a set of defined deliverables, meeting quality standards and requirements. Garies (1991) argues that due to dynamic markets, new developments in environmental and technological arena, companies are facing increasing business complexity. Therefore, companies employ PM to carry out complex and unique tasks. Garies (1991) states that companies resort to PM as a strategy to achieve the following objectives:

■ Organizational flexibility,
■ Decentralization of management responsibility,
■ Concentration on complex problem,
■ Goal-oriented problem-solution processes.

Garies (1991) states that in addition to the traditional industries such as construction and engineering, other industries such as manufacturing, banking, marketing, and tourism are also employing PM as a tool to achieve strategic goals. However, IT, and particularly software development, has been employing PM for many decades. Reiss (1993) finds PM as a human activity that achieves a clear objective against a time scale. Lock (1994) argues that PM has evolved to plan, coordinate, and control complex and diverse activities of commercial and industrial projects. Similarly, Burke (1993) treats PM as a specialized management technique geared to plan and control projects under a strong single point of responsibility. Burke's (1993) definition imparts a key role to the project manager rather than traditional iron triangle of cost, time, and quality. Munns and Bjeirmi (1996) argue that PM is the process of managing the achievement of project objectives by employing the organizational structures, utilizing its resources, and using tools and techniques without disturbing the day-to-day operations of the base organization. Atkinson (1999) refers to British Standard for Project Management, which defines PM as the planning, monitoring, and controlling all aspects of a project and the motivation of all those involved in to achieve project objectives, within the triple constraints of the specified time, allocated cost, and expected quality and performance. Atkinson (1999) favors cost and time as two best guesses for project success, particularly when enough information about the project is not available.

Writing on the important role of PM, Artto and Wikström (2005) state that PM can contribute to the achievement of strategic objectives of organization and businesses. Martinsuo et al. (2006) argue that increased internal complexity and external pressures are the main drivers for the adoption of project-based management. They state that internal complexity is a more compelling driver as compared to external pressures. These authors argue that early adoption of project-based management was intended to manage efficiency and effectiveness of challenges. The results of research indicate that in order to benefit from the project-based management, a firm-wide implementation is a prerequisite. The study also highlights that the benefits of the introduction of project-based management can be in the form of improvements in efficiency and project culture (Martinsuo et al. 2006). PM is not a sub-discipline of engineering anymore, and it can be employed to implement organization strategy, business transformation, and continuous improvement, as well as for product development. Shi (2011) argues that PM is a powerful management solution, and it has become very popular in various industries since 1980s. Shi (2011) states that the value of PM has now been well and truly acknowledged, and proposes that in order to increase value from PM, we must provide the enabling environment. Shi argues that the enabling environments can be created by ensuring that the PM implementation approach should be appropriate and the organizational environment should be favorable for the implementation of PM.

Shenhar and Dvir (2007) state that traditional PM approach is based on the assumption that projects are predictable, fixed, simple, and certain, and the main drivers for traditional PM are triple constraints and one size fits all. These drivers may still be important for a small group of projects, but the traditional model is no longer relevant to the modern-day PM, as today's projects are complex, uncertain, and continuously changing and strongly affected by their environment, technology, and markets. Therefore, Shenhar and Dvir (2007) propose an adaptive PM model. This new model is based on the assumption that projects are not just a group of activities that need to be completed on a given time frame; rather, these are business-related processes that must deliver business results to the organization. Since modern-day projects involve greater level of uncertainty and complexity, such projects should be managed in a more flexible manner (Shenhar and Dvir 2007). Thomas and Mullaly (2007) present a model of PM value, based on 65 case studies conducted on a global scale. This research identifies that there is a high level of correlation between PM practices and project success by achieving project goals and outcomes. Following is the model by Thomas and Mullaly (2007) in which they highlight the environment in which projects exist and are implemented.

Thomas and Mullaly (2007) argue that the decision to implement PM in an organization is guided by the direction of the business of the organization; its focus, vision, and strategic direction; and the business environment in which a given organization is functioning. The PM value model of Thomas and Mullaly is based on three constructs: (1) Organization context: This construct deals with the questions as to what extent the organization does the "right things" in an appropriate manner in the context of PM. (2) Implementation of PM: This construct identifies how the established PM system influences the delivery of projects in an effective and efficient manner.(3) The type of value generated by PM: This construct explores as to what extent the PM capability of the organization influences on the delivery of organizational goals such as cost savings, increased revenues, and what is the return on investment on the PM training by the organization (Thomas and Mullaly 2007). Eskerod and Riis (2009) investigated five case study organizations, where various models of PM have been implemented and identified four value outcomes of these models, namely, efficiency, legitimacy, power and control, and stakeholders' satisfaction. Efficiency value was highly rated, which referred to efficiency gains in cost savings, improved utilization of resources, easier problem solving, and reduction in initial mistakes. Stakeholders' satisfaction was the second most rated value gains of PM models, which deals with meeting clients' expectations, improved processes during project life cycle, and team performance. The legitimacy value includes dealing with customers professionally and good reputation of the organization. The fourth value of power and control includes active involvement of top management, greater transparency, and strong basis for the company to achieve its business objectives.

PMI White Paper (2010) states that now companies recognize the value of time and money investment in PM, as they reap benefits through reduced costs, increased customers, and stakeholders' satisfaction and enhanced competitive advantage. The white paper further states that implementing PM, across the entire organization, enables creating value chain, particularly in high-risk area. It also adds that more than half of the executives informed Economist Intelligent Unit that the role of PM has become more important since 2007 financial crisis. Through effective PM, companies are able to deliver projects on time and on budget, which may ensure more business and help the product hit the market at the most opportune time (PMI 2010). Martinsuo et al. (2012) argue that PM is more than just organizing a temporary organization known as project. PM is also about organizing a parent organization supporting multiple projects. The authors state that an overarching theme in PM is about value generation through projects. Value is not limited to only economic or financial gains but it may also include social, ethical, and ecological value (Martinsuo et al. 2012). Morris (2013) argues that PM should endeavor to create value for the sponsor and other stakeholders by achieving project outcomes. Taking Morris' argument further, Artto et al. (2016) state that the value adding within a project links the front-end PM processes to the back end of the system life cycle such as the operation phase.

PM literature makes a powerful case for PM as the most effective management tool to implement organizational strategy, achieve strategic objectives, and obtain value for the organization. Traditionally, project success debate remained restricted to delivering projects on time, on budget, and meet specifications, but as the discipline matured over the years, more criteria were added, such as customer satisfaction, BR, and value for the organization. There is an increasing emphasis on value generation from PM, and value is not limited to economic and financial gains; in fact, value encompasses social, ethical, and ecological values as well. Various authors have presented a number of PM frameworks such as Shenhar and Dvir (2007) and Thomas and Mullaly (2007) to highlight PM as an effective management tool. The above section highlights the value of the application of PM for organizations. Therefore, it is pertinent to discuss how far PM has been able to help implement the organizational strategy and achieve the strategic objectives through BR.

5.3 Project Success—Benefits Management and Realization

Project success has frequently been discussed over the decades in the PM literature, and achieving project success has been the holy grail of project practitioners and researchers. There are different views on the origins of PM, which according to some, dates back to the construction of pyramids, but Snyder (1987) argues that modern PM appeared as a discipline during the 1950s and Rolstada et al. (2014) state that project success caught the interest of academics in the 1980s. During the 1980s, PM research started investigating project success beyond the "Iron Triangle" of scope, cost, and time, and leading the project success debate. Pinto and Slevin (1988) published a list of 10 project success factors, which is now considered a pioneering work on project success (Rolstada et al. 2014). Since nearly last two decades, the PM success debate has moved from project outputs and is now more focused on project outcomes leading to project BM and BR. Bradley (2010) is credited to have introduced BM initially, and later, he rebranded it as BR. Bareese (2012) highlights that BM and BR have lately received growing attention, and the literature on benefits has been increasing exponentially. In 2009, Association for Project Managers (APM) set up a Special Interest Group (SIG), which has regularly been seeking its member's opinion and releasing survey reports on the significance of BM in organizations of their employment. According to APM (2017) survey report, the members have acknowledged that in their organizations, there is a growing awareness on making BM an integral component of PM, particularly P3M. Bareese (2012) argues that initially BM was employed to evaluate IT investments, but it is universally relevant to other disciplines and professions.

In the PM literature, BR has been hailed as a key success criterion along with the delivery of projects within scope and on time and cost. PM literature has been forcefully arguing for the emphasis on project outcomes rather than project outputs. Mossalaman and Arafa (2016) argue that BR has become a significant factor for projects, and it is common that project success is assessed on the basis of project benefits rather than project completion on time and cost. Researchers have proposed a number of frameworks to effectively package BR into PM processes. Zwikael and Smyrk (2012) argued for an extra fifth phase to the existing project life cycle for project BR. The following section briefly discusses the current research debates on BR and informs how far BR has been adopted by the project practitioners, and it also looks at various propositions by the researchers for effective BR.

Public sector organizations are facing various challenges in their endeavors to implement BRs. In the Australian Public Sector projects, success in BR is patchy and inconsistent. BR scene in Australia is quite similar to other developed countries. Marnewick (2016), who conducted a study of various organizations in South Africa and Holland, states that benefits gained from IT investment projects are not linked to the organizational strategic objectives; therefore, it is certain whether the promised benefits have been harvested or not. Elaborating further research findings, Marnewick (2016) states that organizational culture does not play any role on how BM is practiced in the understudy organizations; however, the adherence to BM practices depends on organizational maturity. The author highlights that organizations do not follow BM best practices, which affects the level of success in the realization of benefits from the investment on information systems projects. These organizations were found aware of the importance of business case and its role in BM. Chih and Zwikael (2015) highlight similar challenges with BM in the Australian public sector organizations. They state that organization do not have the ability to formulate benefits and also do not have processes in place to link the delivered benefits to the promised benefits. Saeed and Abbasi (2019) state that in many Australian public sector organizations, the majority of the project practitioners are aware of the significance of BM and BR. In these organizations, the identification of potential benefits of projects is a requirement for project approval but the commitment to track benefits mellows down during project implementation, and at the project completion, success in BR varies from maximum 50% to embarrassingly nil (Saeed and Abbasi 2019). Chih and Zwikael (2015) suggest that project target benefits must align to organizational goals. These benefits must be measurable and realistic, and should be time specific as well as target value. The authors offer a framework based on six propositions: (1) Project target benefits can be appraised on the basis whether these fit in organizational strategic goals, (2) the employment of a formal benefit formulation process can improve project target benefits, (3) a highly motivated managers will contribute towards improved target benefits, (4) the presence of strong senior executive leadership can contribute towards improved target benefits, (5) the strong executive support in the form of resources allocation, and (6) the prevailing innovative climate provides enabling environment for all the other five constructs.

In order to bridge the disconnect between the organizational strategic objectives and accruing project benefits, Marnewick (2016) suggests a closed loop system, which proposes to link the expected project benefits listed in the business case with the accrued benefits and also connecting these benefits back to the strategic intent. The author states that organizations are aware that the delivery of the promised benefits is a measure of success, but still these organizations do not institute effective processes guaranteeing the delivery of these benefits. The existing BM literature paints varying at times contradicting pictures on the employment of BR in practice. Saeed and Abbasi (2019) state that the majority of the public sector organizations in Australia are aware of the significance of BM, but this awareness rarely translates into positive actions for BR. Marnewick (2016) dispels the common impression found in prevalent PM standards and methodologies that BM should be managed at the project program level. However, Marnewick (2016) states that his research indicates that the organizations do attach benefits to individual projects irrespective of project cost and scope. Therefore, the author proposes that BM should be a part of PM as another knowledge area and traditional PM life cycle be expanded to include benefits delivery and BR. Thus, Marnewick (2016), in fact, echoes the suggestion by Zwikael and Smyrk (2012) in which they have proposed to extend the traditional PM cycle to project BR and suggested that the focus of PM should be shifted from project outputs to project outcomes. Marnewick (2016) also suggests that the focus of project governance should be extended from project delivery to outcomes, thus ensuring BR, which also highlights the call for the inclusion of BR in PM body of knowledge. Zwikael and Smyrk (2012) take the project success debate to a new level by arguing for making project outcomes rather than outputs as the foundation of project evaluation process. The authors endeavor to modify the conventional view from input-process-output (IPO) to input-transform-outcome (ITO) of the project activity.

Saeed and Abbasi (2019) also highlight that project governance can play a critical role in project BR, but in most of the Australian public sector organizations, project governance is more focused on project delivery and BR does not figure in project reports to the project boards and committees.

On the conceptual level, BM and BR have gained currency among the project practitioners as Badewi (2016) states that PM literature describes BM as a mechanism of "initiating, planning, organization, executing, controlling, transitioning, and supporting change" in the organization. Badewi (2016) rejects the commonly held belief among the practitioners that completing project on time and cost will necessarily lead to stakeholders' satisfaction and delivering expected project benefits. His research highlights a strong link between efficient PM ensuring project outputs on time and cost and the desired project BR. Badewi (2016) states that PM practices have a significant impact on project success, and similarly, there is a correlation between the PM and BM. Therefore, the author argues that a combination of PM and BM, implemented through a single governance framework, would increase the potential of success. Serra and Martin (2015) argue that effective BR management helps in highlighting the value and the strategic relevance of projects, which results in effective project governance but also enables the organization to deliver planned benefits through strategic governance. Mossalaman and Arafa (2016) discovered a significant lack of BM at the project level and suggest strong governance in order to effectively implement BM. Similarly, Saeed and Abbasi (2019) state that their study of Australian public sector organizations also highlights that PM processes combined with BM framework ensure a more meaningful project success as BM tools such as BM realization plan, BM profiles, maps, and benefits owners are geared towards projects outputs and outcomes delivery, rather than having (standalone) good outputs or good BM practices only.

5.4 Benefits Management and Realization Frameworks

This section discusses a number of frameworks proposed in the published literature. Most of the frameworks are normative as these have not been empirically applied and tested. However, it would be interesting to discuss these frameworks as these would demonstrate the emerging consensus among the PM experts, on various BM processes. These frameworks provide good insights as to how BR can be implemented at the project, program, and portfolio levels.

As pointed out earlier in this chapter, BM was initially employed by organizations to evaluate investments in information systems (IS)/IT. Therefore, the earlier efforts to develop frameworks for BM also revolved around IS/IT. One such endeavor is the development of the Active Benefits Realization (ABR) framework. The ABR approach is based on the notion of continuous participative evaluation and contingency philosophy. By contingency philosophy, the authors mean that the information systems, development activities, and stakeholders' roles are dynamic and these continuously evolve. ABR can be employed for the evaluation of even existing information systems (Remenyi and Sherwood-Smith 1998).

Like other BR frameworks, ABR requires active participation from stakeholders; therefore, it entails agreed roles and responsibilities. The authors state that in ABR, the principal stakeholders are identified at the outset and they agree to remain involved in the entire process. Remenyi and Sherwood-Smith (1998) identify the following key stakeholders in the ABR process, though there are other secondary groups of stakeholders:

- Line managers and end users of the system,
- Accountants and financial officer,
- Information system people.

ABR is a participative and reiterative process, which ensures that project is on the right track for BR. The reiteration process comprises the following seven activities (Figure 5.1):

- Initiation of the project,
- Production of pictures,
- Agreement to proceed,
- System development,
- Evidence collection,
- Review and learning,
- Update of pictures.

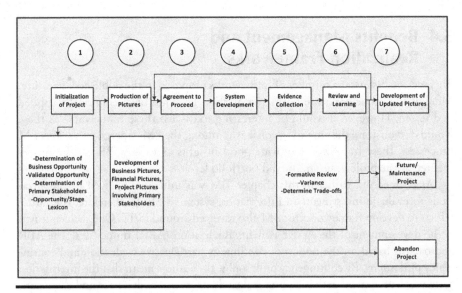

Figure 5.1 ABR process (Remenyi and Sherwood-Smith 1998) adapted from Viklund and Tjernstrom (2008).

As an initial attempt to develop a benefits evaluation process, the ABR is a valuable framework specifically designed for the evaluation of investment in information systems. Like various other frameworks developed chronologically later, ABR emphasizes an active participation of the key stakeholders for the success of the benefits evaluation process.

According to Viklund and Tjernstrom (2008), the BM Framework of the Office of Government Commerce is based on the presumption that BM ensures that the target business change or policy outcomes are defined and measurable to make a convincing justification for investment and also ensures that the change is ultimately achieved. The OGC BM framework is based on four processes, namely, benefits identification, benefits optimization, tracking and BR, and finally reviewing and maximizing benefits. This process continues throughout the life cycle of the program, and it terminates with the formal closure of the program. With the closure of the program, operations managers assume responsibilities to monitor and optimize planned benefits. Like others, the OGC framework also puts great emphasis on the role and responsibilities; therefore, with the closure of program, managers play a key role in BR. The framework identifies the following important roles in the BM process:

- Senior responsible owner,
- Program manager,
- Program office,
- Project manager,
- Assurance and validation.

The OGC framework calls for a BM strategy, which is not found in other frameworks, though various other frameworks play a key role to a BM plan. Like all other frameworks such as Bradley (2010) and Ward and Daniel (2012), the OGC BM framework also offers four quite similar iterative processes, namely, benefits identification, benefits optimization, realizing and tracking, and reviewing and maximizing. In this framework, BM strategy, BM realization plan, benefits identification, and benefits optimization relate to project initiation and planning phases, whereas the last two processes, namely, realization and tracking, and reviewing and maximizing, belong to post-implementation phases, which again is similar to other frameworks.

Bradley (2010) proposes BR management process based on six phases, which can be applied to programs and projects, which are as follows:

- Set vision and objectives,
- Identify benefits and changes,
- Define initiatives,
- Optimize initiatives,
- Manage initiatives,
- Manage performance.

Bradley's (2010) framework is quite similar to the Office of Government Commerce framework for BM, which comprises four phases. However, compared to all other frameworks discussed here, Bradley awards a central role to stakeholders in most of the stages, though in other frameworks, the role of stakeholders is equally important but that is implicit rather than explicit as in Bradley.

Ward and Daniel (2012) developed a framework, which is also known as the Cranfield Process, after intensive research on organizations, who successfully employed it. Ward and Daniel (2012) claim that the feedback from organizations under the study has confirmed that this framework is applicable not only to IS/IT projects but also equally effective in other initiatives for business development and change programs. Ward and Daniel (2012) argue that BM is not a standalone process; rather, there is an interdependency of changing organizational processes, relationships, roles, and working practices inside and at times outside the organization. Ward and Daniel (2012) see a direct interdependency between change management and BR, which means that effective change management is a key precursor to successful BR.

Ward and Daniel (2012) state that BM framework is iterative in nature, and it comprises five phases, which are as follows:

■ Identify and structure benefits,
■ Plan business realization,
■ Execute benefits plan,
■ Review and evaluate results,
■ Establish potential for further benefits.

In Ward and Daniel's (2012) framework, the initial three phases relate to initiation and execution and the last two phases are concerned with the post-implementation stage. The authors claim that this framework is being successfully employed by over 100 major organizations in UK, the United States, and Australia, for effective BR.

Chih and Zwikael (2015) propose a conceptual framework for target benefit formulation and propositions. This framework comprises strategic fit, target value, measurability, realism, target date, accountability, and comprehensiveness.

The NSW Government (2015) developed a framework with a purpose to introduce a standard approach to BM within its agencies. This framework is intended to help strategy groups, operational business areas, program and project teams, individuals, and business benefits owners. This framework is also expected to help a Project Management Office (PMO) to enhance decision making and improve change management. This framework is based on 15 principles and 4 processes, which have been developed on the basis of best practices and current experiences across New South Wales (NSW) agencies (NSW Government 2015). Following are the principles of this framework:

■ Business needs to be understood first as outcomes,
■ Benefits must be aligned to the organization's strategic goals,

- BR is an end-to-end process covering the investment life cycle,
- BM is the foundation of a successful business case,
- Change programs that deliver benefits,
- Benefits are not automatic,
- Benefits can be financial and non-financial,
- Intermediate outcomes lead to realize final benefits,
- Benefits must be quantifiable,
- The business must own the benefits,
- The number of expected benefits should be manageable,
- BM must be linked with project/program management,
- BR must be integrated with other relevant processes,
- Benefits must be communicated.

Similarly, the New Zealand Government (2016) identifies a framework based on four phases, namely, identification, analysis, planning, and realization and reporting. This framework is aligned to investment cycle which is also based on the four linear phases, namely, thinking phase, planning phase, doing phase, and reviewing phase.

Peppard et al. (2008) recommend a Benefits Dependency Network (BDN) approach for effective BM. This approach calls for a bespoke BDN to guide investment decision making particularly in IT. Under the BDN approach, questions are asked if the existing problem is resolved with the proposed investment, then what would be the target improvements. Then, the identified improvements become objectives for the investment, which will lead to the identification of expected benefits, and metrics will be defined to measure the expected benefits. The authors claim that constructing a BDN can save organizations from making bad investments. BDN approach has been successfully employed by over a hundred organizations of various sizes, in the public and private sectors across the world. The BDN has been adopted as "best practice framework" by the State of Queensland in Australia for IT investments and is being introduced by a government in Europe to improve the management of IT investments (Peppard et al. 2008). However, the authors did not say whether the BDN framework can be applied to investments other than IT.

5.5 A Focus on PM in the Australian Public Sector Organizations

PM in the Australian public sector poses specific challenges to project practitioners as according to Patanakul et al. (2016), these projects are implemented in a complex and political environment, face a lot of public scrutiny, meet strict compliance requirements, and involve multiple layers of stakeholders. A number of research studies highlight that in the Australian public sector at both the Commonwealth and States levels, projects struggled to achieve organizational strategic objectives because either the projects were not properly aligned to strategic objectives or the objectives did not remain stable during the life cycle of these projects. Young et al. (2012) studied the State of Victoria, public sector investment of around $30 billion between 1999–2009–2010, in various public sector agencies to achieve strategic goals through projects. They investigated the realization of strategic goals through the implementation of projects and rate the PM practices and framework in the State of Victoria better than the best practices. However, they argue that despite being better than the best practice in the public sector PM, there is no evidence that projects have contributed to the improvement of strategic goals in any agency under study in Victoria. To the authors, this is a matter of concern, as projects are generally considered enablers of strategies. Young et al. (2012) state that irrespective of the success or failure, the contribution of projects towards the organization strategy is negligible in two major public sector agencies of the Victorian government. Therefore, the authors think that problem may lie in how the projects are selected and governed.

Similarly, Young and Grant (2015) conducted another study of public sector investment in the State of NSW, where the authors investigated public sector agencies' investment of Aus$100 billion over a period of ten years to achieve strategic goals through project implementation. Although the variables in this study were almost similar to the research in Victoria in 2012, the authors argue that whole of government (WG) environment was the new factor that may have instigated the slight difference of conclusions between these two studies. Young and Grant (2015) argue that before the adoption of WG in the NSW, the strategic objectives were unstable and 30% of strategic objectives would disappear every year. Therefore, the authors state that projects are twice as effective to contribute to strategic goals in the presence of stable strategic priorities and centralized project oversight. The authors believe that due to WG leading to the establishment of PMO, the stability of the strategic goals could be one of the outcomes of this change. Concluding their research of the NSW public sector projects, Young and Grant (2015) argue that projects contributed little to achieving strategic goals, as only 20% of the projects were able to realize strategic goals. This conclusion signifies improvement though to a small degree, as compared to their conclusions of Young et al. (2012), who conclude that there is no evidence that projects contribute to the realization of strategic goals.

As highlighted above that the projects failed to achieve strategic objectives in Victoria and NSW, due to changing strategic objectives Patanakul et al. (2016) also problematize the success criteria in public sector projects in UK, the United States, and Australia. Writing about the performance evaluation of these projects, Patanakul et al. (2016) state that the audit offices of the respective countries evaluated projects on the basis of the traditional criteria of time, cost, and scope rather than the BR from the projects. Therefore, Patanakul et al. (2016) argue that the evaluation of project success cannot be based only on efficiency variables such as cost and time; rather, the authors argue for the inclusion of benefits realized from the project as criteria of success. The authors discovered that failure to spell out expected project benefits resulted in the failure of projects to provide intended outcomes. Therefore, Patanakul et al. (2016) argue that the focus of PM on BR can lead to enhanced project performance. For this to happen, the authors argue for specific and definable target benefits and a comprehensive methodology for evaluating target benefits. The authors state that in addition to focus on project cost, time, scope, and quality, the projects should also align to agency's strategy, which is critical for project success. However, a recent study by Saeed and Abbasi (2019) highlights that there is a widespread awareness on the importance of BM and BR in the Commonwealth public sector organizations. This research finds that project benefits are identified as a requirement for the project approval. But as the project implementation starts, the focus on outputs delivery overwhelms and the commitment to benefits tracking and BR loses its steam, at times to the extent that it is ignored. However, there are some pockets of excellence, where BR has been taken very seriously and encouraging results have been achieved. Among the factors that lead to the loss of focus on BR are the lack of skills, time, and budget as well as poor governance for BR. This research also highlights that BR can be taken seriously only when the push comes from the top management and project governance directs it focus to BR, in addition to progress on deliverable, cost, and schedule (Saeed and Abbasi, 2019).

This brief review of the PM practices in the Australian public sector highlights that future projects need to be managed more effectively, wherein the focus on project outputs and outcomes should be assigned similar importance, so that the hefty investments of tax payers' money should result in expected benefits. The lessons learned through public sector projects so far should inform how the future projects, particularly cybersecurity projects, should provide the expected security to the Australian public and businesses. The following section discusses the current state of cybersecurity projects.

5.6 Cybersecurity Projects

Cybersecurity projects fall into IT projects, and these projects would face the similar challenges, which all other IT-related projects encounter. These challenges are delayed completion, cost overruns, scope creep, expected benefits not realized, and organizational strategic objectives not achieved. Thomas and Fernández (2008) conducted a study of IT project success in 36 companies of three sectors, namely, mining, finance and insurance, and utilities (power, gas, water). The authors state that project success in IT projects is as elusive as it is in other projects in various sectors. According to Standish Group CHAOS Report 2017, about $250 billion was spent on 175,000 projects of IT applications and only 28% of software projects at small-sized companies were completed on time and cost. Similarly, 31% of projects will be canceled and nearly 53% projects will be 189% over budget. According to the report, in the United States, the project success rate in the medium-sized companies was 16% and larger-sized companies present a dismal situation in project completion, which is mere 9%. The report did not highlight whether the company size has any impact on project success rate. Similarly, according to Ismail (2017), in UK alone 37 billion pounds was wasted on failed agile IT projects.

Katharina (2018a) argues that a solid PM practice is a prerequisite for cybersecurity projects to be successful. She identifies the following five benefits of a PM process in cybersecurity projects:

- Streamline project execution,
- Strategic alignment,
- Optimized resource allocation,
- Continuous improvement,
- Problem resolution and risk management.

Katharina further states that managing cybersecurity of the organization turns out to be a project in itself. With PM in place, the executives appreciate that cybersecurity projects are aligned with the overall business strategy (Katharina 2018b). Aligning projects with the organizational strategic objectives is one of the major challenges in the private and the public sector organizations. Cybersecurity projects in the Australian public sector are expected to suffer from the same challenges which other projects and particularly IT projects currently face. These projects may not succeed in achieving organizational objectives of securing public data, privacy of public information, and the security of defense-related information. As mentioned in the introduction, cyber breaches have already taken place in the leading Australian universities and Australian Federal Parliament. Olsen (2014) argues for the adoption of a PM approach for the implementation of cybersecurity projects. A PM approach for cybersecurity projects will ensure that project outcomes align with the organizational strategic goals. A PM approach will add a degree of formality and standardization to project initiatives for a greater project success (Olsen 2014). Therefore, it is pertinent that project practitioners in the Australian public sector should direct their efforts to BR of investments made into cybersecurity through projects. The following section makes useful recommendations for ensuring greater and real success through projects' BR.

5.6.1 Making Cybersecurity Projects Successful for Benefits Realization

Like all other projects in the public sector, cybersecurity projects are expected to face similar challenges, discussed in the literature review above. These problems are a low rate of project success not only on the basis of scope, cost, and time but also on failure to harvest the expected benefits from investments made in cybersecurity projects, as there is an increasing demand for judging projects on the basis of outcomes and benefits rather than outputs. This section makes a number of recommendations for improvements in the current PM practices for enhanced success on the basis of a recent research conducted for the Doctorate of Project Management in the Australian public sector organizations. The following section briefly touches upon various problem areas on the basis of initial findings of this research.

Project governance has been identified as the most key player in ensuring project success, particularly for BR. Bekker and Steyn (2008) state that on the lines of the corporate governance, the concept of governance has gained acceptance in various other areas, including PM. The research study mentioned above highlights the important role of project governance in an effective BR. But most of the participants, who were interviewed for this research, informed that the lackluster support from the project governance in most of the case study organizations except one is hemorrhaging BR. In some cases, the executives on project boards and committees are either not fully aware of their roles and responsibilities or are focused only on project deliverables, schedule, and costs. This research also highlights that project progress reports do not dilate on BR, as there is no reporting requirement on BR. One informant said, "They [project board] like when you provide benefits realization plan but if you do not, no body loses their sleep." Another informant commented, "you may find very professional people sitting in the project board but at times, they just sit in the meetings and play with their mobile phones during the meetings." Most of the participants are of the opinion that BR can be taken seriously by the program and project managers, only if the project governance shows its commitment to it. One participant said, "despite a lot of time of project managers is spent on reporting but benefits do not figure in progress reports to project committees, as mostly these reports focus on cost and time." "If the project governance makes reporting on benefits tracking mandatory, the project managers will do it, in no time." Another participant, who is personally committed to the cause of BR, said, "people in [project] governance do not understand benefits, they look at it with a scatter-gun approach and do not see at BM as a process from start to the end." Therefore, it is recommended that cybersecurity projects governance boards should comprise people who think beyond deliverable, cost, and time, and are genuinely committed to and trained in BR.

Most of the researchers have emphasized the role of benefits owner in effective benefits harvesting. Peppard et al. (2008) argue that benefits owners should be nominated and the responsibility for the realization of each benefits must be assigned to benefits owners. Similarly, Zwikael and Smyrk (2012) state that as the project manager is responsible for the delivery of project outputs, project owner should be accountable for project BR. For this to happen, the authors proposed to extend project life cycle to project realization and measurement.

An overwhelming majority of the research participants agreed on the key role of benefits owners in BR and also suggested that the benefits owners should be nominated from among the operation managers/business managers, whose departments are expected to use the service or the product. However, the identification of benefits manager/owner necessitates the involvement of such a person during the development of BR plan. But in one case study organization, where billions is spent on the maintenance and development of new IT applications, there exists a lack of meaningful engagement between the business and operations departments, which has been highlighted and ultimately seriously impacts BR. In another case study organization, a participant said, "for effective benefits realization, benefits owners need funding, human resources and skills, which are not provided at the moment."

As discussed above, various researchers have proposed a number of frameworks for BM suggesting that these frameworks should be implemented in conjunction with the existing PM methodologies (Zwikael and Smyrk 2012; Chih and Zwikael 2015). Effective BM and BR is possible through BM frameworks, and various case study organizations acknowledged the existence of frameworks. But mere having frameworks does not make BR happen automatically. In the studied organizations, benefits are managed through PM methodologies, specific BM frameworks, and at times under risk and quality management frameworks. In some organizations, BM frameworks have been made available but the use varies from program to program. The use of BM frameworks also depends on the discretion of project and program managers, as some project managers take BM more seriously than others. The BM framework should be used in conjunction with the PM methodologies such as Managing Successful Programs (MSP).

Digital Transformation Agency (DTA), which provides advice to government and its agencies on digital investments, should be given more active role in Information & Communication Technology (ICT) projects and DTA should actively pursue BM and BR on all new projects and particularly cybersecurity projects. At the moment, DTA has a mere consulting and expert role, and its involvement depends on the host department. According to one informant of DTA, the Department of Defence has been most actively seeking guidance from DTA on its various ICT projects.

PMO has a special role to play in improving overall PM practices and particularly BM and BR of projects. Most of the project practitioners supported an active role of PMOs in their organizations but so far, they are dissatisfied with the performance of PMOs. According to research participants, PMOs should help program and project managers in developing BR plans and seek sustained support from project governance. The active role of PMO in BR has been demonstrated in a case study which has been very successful in BR. In this case study organization, program and project managers, PMO team, and executives in governance roles were provided training in BM. This organization has been able to harvest benefits up to 50% during last 18 months since the program started three years ago. One of the informants in this organization said, "We have 50 percent success rate in benefits harvesting and now we are getting to point where we are quantifying and harvesting [benefits] and we are starting to get to the point, where we are going back [to the benefits owner] and saying okay now where are your benefits." This case study also highlights that the support of top management is the key to successful implementation of BR, as the push for BM has come from the relevant minister and the message has been re-emphasized by the top executive through project governance requiring reporting on benefits tracking during and post-project implementation.

The main challenge facing BM in the Australian public sector is not the unavailability of the BM framework but rather integrating the BM framework into PM processes and continuous commitment to BR by the senior executives. Various organizations researched by Saeed and Abbasi (2019) had frameworks available but the use has been left at the discretion of the program managers, because there was no compulsory requirement form the top management or the project governance to realize benefits using a particular framework. As we have discussed above, some of the BM frameworks are simple (NZ Government 2016), whereas others are complex (Remenyi and Sherwood-Smith 1998); organizations can start with a simple framework, and as maturity is achieved, more complex frameworks can be gradually adopted. Most of the practitioners interviewed by Saeed and Abbasi (2019) acknowledged that BR can be adopted only if the top management demonstrates its commitment and seriousness. The program managers and the project governance need a change of mindset, and while focusing on project delivery on time and cost, BR should be imparted due importance as well. It should be mandatory to report on benefits tracking during the project implementation phase and benefits harvesting once the project service or product becomes operational. For effective BR, the benefits owner/capability manager needs to be identified at the beginning and she or he should be responsible to report progress on BR after implementation. However, benefits owners should be provided with necessary resources in the form of funds, time, and training, so that BR can be conducted in a meaningful manner. Cybersecurity projects can turn out to be a great opportunity to take one small step towards a great leap in the history of PM.

5.7 Future Works and Conclusion

This chapter has discussed the significance of cybersecurity threats to governments and businesses all over the world. Governments and businesses are fast realizing the need for building defenses against cyber threats. New investments in cybersecurity are expected to be made through projects. Therefore, this chapter has discussed the key role of projects in the global economy, value creation, and BR. This chapter highlights the need for learning from failed projects, particularly in the public sector, where organizations spent millions without achieving strategic goals. Therefore, this chapter argues for employing BM and BR perspective to all cybersecurity projects by linking the organizational strategic goals to projects and paying special attention to harvesting the benefits of investments.

References

APM (2009) Benefits Management, "A strategic business skill for all seasons", www. apm.org.uk/news/benefits-management-a-strategic-business-skill-for-all-seasons/, accessed on 18 Aug, 2017.

Artto, K., T. Ahola, and V. Vartiainen (2016). "From the front end of projects to the back end of operations: Managing projects for value creation throughout the system life-cycle." *International Journal of Project Management* **34**(2): 258–270.

Artto, K. A. and K. Wikström (2005). "What is a project business?" *International Journal of Project Management* **23**(5): 343–353.

Atkinson, R. (1999). "Project management: Cost, time and quality, two best guesses and a phenomenon, it's time to accept other success criteria." *International Journal of Project Management* **17**(6): 337–342.

Australian Broadcasting Corporation (20 Feb, 2019). The Cyber-Attack on Parliament Was Done By a 'State Actor'—Here Is How Experts Figure that Out, www.abc.net.au/news/2019-02-20/cyber-activists-or-state-actor-attack-how-experts-tell/10825466, accessed on 27 July, 2019.

Australian Cyber Security Centre (2017). Threat Report, www.cyber.gov.au/publications/acsc-threat-report-2017, accessed on 26 July, 2017.

Badewi, A. (2016). "The impact of project management and benefits management practices on project success: Towards developing a project benefits governance framework." *International Journal of Project Management*, **34**(4).

Bareese, R. (2012). "Benefits realization management: Panacea or false dawn." *International Journal of Project Management*, **30**(3): 341–351.

Bekker, M. C., and H. de V Steyn (2008). The Impact of Project Governance Principles on Project Performance, *Portland International Conference on Management of Engineering and Technology*, 27–31 July, 2008.

Bradley, G. (2010). *Fundamentals of Benefits Realization*, TSO, Belfast.

Bredillet, C. N., K. Conboy, P. Davidson, and D. Walker (2013). "The getting of wisdom: The future of PM university education in Australia." *International Journal of Project Management* **31**(8): 1072–1088.

Burke R. (1993). *Project Management*, John Wiley and Sons, Chichester, as cited in Atkinson, R. (1999). "Project management: cost, time and quality, two best guesses and phenomenon, it's time to accept other success criteria." *International Journal of Project Management*, **17**(6): 337–342.

Chih, Y. Y. and O. Zwikael (2015). "Project benefit management: A conceptual framework of target benefit formulation." *International Journal of Project Management*, **33**(2): 352–362.

Digital Transformation Agency, www.dta.gov.au/, accessed on 23 July, 2019.

Eskerod, P. and E. Riis (2009). "Project management models as value creators." *Project Management Journal* **40**(1): 4–18.

Garies, R. (1991). "Management by projects: The management strategy of the 'new' project management company." *International Journal of Project Management*, **9**(2): 71–76.

I. C. B. (ND). www.aipm.com.au, accessed on 21 Nov, 2016.

Ismail, N. (2017). UK Wasting £37 Billion a Year on Failed Agile IT Projects, www.information-age.com/uk-wasting-37-billion-year-failed-agile-it-projects-123466089/, accessed on 15 Jul, 2019.

Katharina, G. (2018a). Why Project Management is Essential For Successful Cybersecurity Projects, www.hitachi-systems-security.com/blog/5-benefits-of-project-management-for-cybersecurity/, accessed on 14 July, 2019.

Katharina, G. (2018b). How to Align Your Security Strategy with Your Business Goals, www.hitachi-systems-security.com/blog/how-to-align-your-security-strategy-with-your-business-goals/, accessed on July 17, 2019.

Landis-Handley, J. (June 5, 2016). Why the ANU Was the Target of a Massive Cyber Attack, www.crikey.com.au/2019/06/05/cyber-attack-data-breach-anu/, accessed on 27 July, 2019.

Lock, D. (1994). As cited in Atkinson, R. (1999). "Project management: Cost, time and quality, two best guesses and phenomenon, it's time to accept other success criteria." *International Journal of Project Management* **17**(6): 337–342.

Marnewick, C. (2016). "Benefits of information system projects: The tale of two countries." *International Journal of Project Management* **34**(4): 748–760.

Martinsuo, M., H. G. Gemünden, and M. Huemann (2012). "Toward strategic value from projects." *International Journal of Project Management* **30**(6): 637–638.

Martinsuo, M., N. Hensman, K. Artto, and A. Jaafri (2006). "Project based management as an organizational innovation: Drivers, changes, and benefits of adopting project based management." *Journal of Project Management* **36**(3): 457–461.

Morris, P. W. (2013). *Reconstructing Project Management*, John Wiley & Sons, New Jersey.

Mossalaman, A. and M. Arafa (2016). "The role of project manager in benefits realization management as a project constraint/driver." *Housing and Building Research Centre Journal* **12**(3), www.sciencedirect.com/science/article/pii/S1687404815000103, access on 10 July, 2017.

Munns, A. K. and B. F. Bjeirmi (1996). "The role of project management in achieving project success." *International Journal of Project Management* **14**(2): 81–87.

NSW Government (2015). BRM Framework, www.finance.nsw.gov.au/publication-and-resources/benefits-realisation-management-framework, accessed on 27 Aug, 2017.

NZ Government (2016). Managing Benefits from Projects and Programs: Guide for Practitioners, www.treasury.govt.nz/statesector/investmentmanagement/plan/benefits/guidance/managingbenefits-guidance.pdf, accessed on 20 Aug, 2017.

Olsen, B. (2014). The Role of Projects and Programs Management in Cyber Security, www.linkedin.com/pulse/20140924122921-6445912-the-role-of-project-and-program-management-in-cyber-security/, accessed on 27 July, 2019.

Patanakul, P., Y. H. Kwak, O. Zwikael, and M. Liu (2016). "What impacts the performance of large-scale government projects?" *International Journal of Project Management* **34**: 452–466.

Peppard, J., J. Ward, and E. Daniel (2008). "Managing the realization of business benefits from IT investments." *MIS Quarterly Executive* **34**(1): 1–11.

Pinto, J. K. and D. P. Slevin (1988). Critical Success Factors Across the Project Life Cycle, Project Management Institute.

PMI White Paper (2010). The Value of Project Management, www.pmi.org/-/media/pmi/documents/public/pdf/white-papers/value-of-project-management.pdf, accessed on Jan 4, 2017.

Oisen, R. P. (1971). As cited in Atkinson, R. (1999). "Project management: Cost, time and quality, two best guesses and phenomenon, it's time to accept other success criteria." *International Journal of Project Management* **17**(6): 337–342.

Reiss, B. (1993). Project Management Demystified. E and FN Spon, London, 1993, as cited in Atkinson, R. (1999). "Project management: Cost, time and quality, two best guesses and phenomenon, it's time to accept other success criteria." *International Journal of Project Management* **17**(6).

Rolstada, A., I. Tommelein, P. M. Schiefloe, and G. Ballard (2014). "Understanding project success through analysis of project management approach." *International Journal of Projects in Business* **7**(4): 638–660.

Saeed, M. A. and A. Abbasi (2019). Project Benefits Realization - Academics Dream or Practitioners' Nightmare, Project and Program Management Symposium, UNSW Canberra.

Serra, C. E. M. and K. Martin (2015). "Benefits realization management and its influence on project success and on the execution of business strategies." *International Journal of Projects in Business* **33**(1): 1–14.

Shenhar, A. J. and D. Dvir (2007). *Reinventing Project Management: The Diamond Approach to Successful Growth and Innovation*, Harvard Business Review Press, Brighton, MA.

Shi, Q. (2011). "Ret hinking the implementation of project management: A value adding path map approach." *International Journal of Project Management* **29**(3): 295–302.

Snyder, J. R. (1987). www.pmi.org/learning/library/modern-project-management-disciplines-direction-1810, accessed on Dec 29, 2016.

Standish Group (2017). CHAOS Report, accessed on July 15, 2019. www.standishgroup.com/store/services/10-chaos-report-decision-latency-theory-2018-package.html.

Telstra security Report (2019). www.telstra.com.au/business-enterprise/news-research/security/research/security-report-2019?gclid=EAIaIQobChMI0fOHz4274wIVlIRwCh3iJgO7EAAYAyAAEgI4vvD_BwE&gclsrc=aw.ds, accessed on 17 July, 2019.

The Guardian (June 6, 2019). China Behind Massive Australian National University Hack, Intelligence Officials Say, www.theguardian.com/australia-news/2019/jun/06/china-behind-massive-australian-national-university-hack-intelligence-officials-say, accessed on 26 July, 2019.

Thomas, G. and W. Fernández (2008). "Success in IT projects: A matter of definition?" *International Journal of Project Management* **26**(7): 733–742.

Thomas, J. and M. Mullaly (2007). "Understanding the value of project management: First steps on an international investigation in search of value." *Project Management Journal* **38**(3): 74–89.

Viklund, K. and V. Tjernstrom (2008) Benefits Management and it applicability in practice, unpublished thesis, IT University of Goetborg, https://gupea.ub.gu.se/bitstream/2077/10452/1/gupea_2077_10452_1.pdf, accessed on Feb 10, 2018.

Ward, J. and F. Daniel (2012) *Benefits Management: How to Increase the Business Value of Your IT Projects*, John Wiley and Sons, New Jersey.

Young, R. and J. Grant (2015). "Is strategy implemented by projects? Disturbing evidence in the State of NSW." *International Journal of Project Management* **33**(1): 15–28.

Young, R., M. Young, E. Jordan, and P. O'Connor (2012). "Is strategy being implemented through projects? Contrary evidence from a leader in New Public Management." *International Journal of Project Management* **30**(8): 887–900.

Zwikael, O. and J. Smyrk (2012). "A general framework for gauging the performance of initiatives to enhance organizational value." *British Journal of Management* **23**(S1): 6–22.

ARTIFICIAL INTELLIGENCE IN CYBER SECURITY

Chapter 6

An Investigation of Performance Analysis of Machine Learning-Based Techniques for Network Anomaly Detection

Kevin Chong, Mohiuddin Ahmed,
and Syed Mohammed Shamsul Islam
Edith Cowan University

Contents

6.1 Introduction

Public facing networks of today are under constant threats. Cyber criminals are actively probing and attacking any network of significance. According to Cisco, 31% of organizations have experienced cyber-attacks on operational technology infrastructure, while Accenture reports from statistics collected in 2017 that there are over 130 large-scale targeted breaches in the United States per year, and these numbers are on the rise (Sobers, 2019). Many of these attacks are now sophisticated with the usage of artificial intelligence (AI) (Dutt, 2018; Elazari, 2017). Deployment of effective network intrusion detection systems (NIDS) to combat against attacks at such an unprecedented scale is paramount.

Machine learning (ML)-based NIDSs have shown potential in detecting such sophisticated attacks (Niyaz, Sun, Javaid & Alam, 2016; Tang, Mhamdi, McLernon, Zaidi, & Ghogho, 2016). With adequate training and correct algorithm, deep learning-based and ML-based techniques are known to detect dynamic zero-day attacks where static rule-based systems could not. Although showing potential, AI-driven cyber security is still in its infancy, necessitating much more study and research.

The purpose of this study is to apply deep learning and ML techniques to detecting network intrusions. The paper starts off in the "2 Preliminary Concepts" section defining the algorithms chosen in this research together with their justification. The "3 Literature

Review" section references related works and explains how this study extends on what had been worked on previously. Next in the "4 Research Methodology" section, it discusses the UNSW-NB15 dataset ("The UNSW-NB15 Dataset Description," 2018) used in this study, followed by pre-processing of data, performance evaluation metrics, and our implementation setup. The results from our experiments are then presented in the following "5 Results" section. The final "6 Discussions" section discusses the insights we had derived from our experimental results.

6.2 Preliminary Concepts

This research looks into two algorithms: Random Forests (RF) and deep neural network (DNN). RF is an ensemble of decision tree models, which are statistical models, while DNNs are neural network-based models where weights and biases are adjusted via iterations of training. Due to their considerable differences in architecture, these algorithms were chosen so we can contrast how two vastly different algorithms perform given the same dataset.

6.2.1 Random Forests

RF is a general-purpose machine learning (ML) model proved to be effective across various classification and regression problems. In contrast to single decision tree models, a RF model's effectiveness lies in training up numerous random and uncorrelated decision trees, by which each of these trees will be consulted as part of the process in forming the final prediction. This is what is known as an ensemble learning algorithm, an algorithm that uses multiple learning algorithms to obtain a better predictive result than what could be obtained from any of the constituent algorithm alone ("Ensemble learning," 2019). Figure 6.1 depicts the architecture of a RF model.

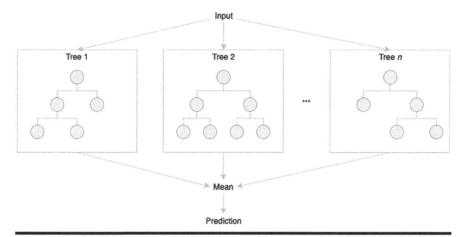

Figure 6.1   Architecture of RF.

Additionally, given the right number of trees and features, the RF algorithm can overcome the issue of overfitting. Single tree models often have high variance in their predictions, and this problem is addressed in RF by averaging the results from across all the random trees.

RF also has a feature which allows the importance of each feature to be ranked against others. Having such an indicator gives us great insight into the UNSW-NB15 dataset and hence, also the problem we have at hand.

6.2.2 Deep Neural Network

A DNN model is composed of multiple layers sitting in between the input layer and the output layer. A record is passed into the DNN at the input layer, this layer performs a set of operations and produces an output, and the output is then being passed as input to the next layer. This process continues through all the hidden layers until it gets to the output layer, where the final prediction is produced. Figure 6.2 depicts the architecture of a DNN model.

One of the strengths of a DNN model lies in its ability to model complex non-linear relationships. The multiple hidden layers in a DNN architecture allow composition of features from lower layers, allowing it to model complex relationships with fewer nodes compared to a shallow network ("Deep learning," 2019).

DNNs have shown good results against the KDDCUP 99 ("KDD Cup 1999 Data," 1999) and NSLKDD datasets in previous works (Yin, Zhu, Fei, & He, 2017). It would also be interesting to see how well DNN scores against the UNSW-NB15 dataset.

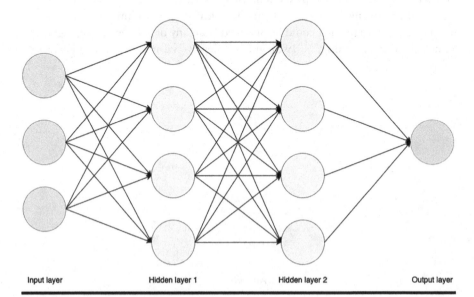

| Input layer | Hidden layer 1 | Hidden layer 2 | Output layer |

Figure 6.2   Architecture of DNN.

6.3 Literature Review

The research conducted by Niyaz, Sun, Javaid, and Alam focuses on using self-taught learning technique based on sparse autoencoder and soft-max regression to implement an NIDS. The conclusion from this work is that the proposed NIDS performed very well compared to other implementations for anomaly traffic detection using the same dataset (Niyaz, Sun, Javaid & Alam, 2016).

Another study focused on using deep learning approach in software-defined networking for flow-based detection. This research concludes that even though the results were not good enough for commercial use, it did, however, show real potential for using deep learning technique for flow-based anomaly detection systems (Tang, Mhamdi, McLernon, Zaidi, & Ghogho, 2016).

The possibility of applying recurrent neural network (RNN) to network intrusion detection was being explored in another study. This particular RNN consists of a hidden layer where the output of the nodes in this layer is fed back to the same hidden layer. The authors of this paper remarked that not only does their system have a strong modeling ability for intrusion detection, it also has high accuracy in its classification results (Yin, Zhu, Fei, & He, 2017).

Another previous work compared a number of classification techniques for network intrusion detection, such as Naïve Bayes Tree, Best-First Tree, J48, Random Forests Tree, and Multi-layer Perceptions. The finding in this research led the researchers to favor Naïve Bayes and Best-First Tree classifiers for the purpose of anomaly network traffic detection (Aziz, Hanafi, & Hassanien, 2017).

While some of the results from these previous works were promising, in this research we want to focus our work on ML-based technique in comparison with the more specific deep learning-based technique. Additionally, our work was conducted using the UNSW-NB15 dataset, while all the above researchers used the older NSLKDD ("NSL-KDD dataset," n.d.) dataset.

6.4 Research Methodology

6.4.1 Dataset

Before the UNSW-NB15 dataset, its creators realized that there was a lack of comprehensive dataset suitable for research in the area of NIDS. The KDDCUP 99 and NSLKDD datasets were over a decade old at that time which lacked modern network traffic, especially those found in the low foot print attack environment. Furthermore, the KDDCUP 99 training set was criticized for not being balanced with too many redundant and missing records. The aim of UNSW-NB15 dataset was to address the shortcomings mentioned above (Moustafa & Slay, 2015a,b; Moustafa & Slay, 2016). For these reasons, the UNSW-NB15 dataset was chosen for this study.

6.4.1.1 Dataset Labels

The training set has 175,341 records, while the testing set has 82,332 records (note that the dataset files downloaded from the original source were named incorrectly: the training set file with 175,341 records was incorrectly named "UNSW_NB15_testing-set.csv," while the testing set file with 82,332 records was incorrectly named "UNSW_NB15_training-set.csv"). Each record is either a normal-type record or an attack-type record. Additionally, if a record is labeled as attack, it is labeled with one of the following nine attack categories: Fuzzers, Analysis, Backdoors, DoS, Exploits, Generic, Reconnaissance, Shellcode, or Worms. The description for each attack category and the number of records found in the two datasets are tabulated in Table 6.1.

Table 6.1   Description and Number of Records for Each Attack Category

Attack Category	Description	Number Records in Training Dataset	Number Records in Testing Dataset
Normal	Natural transaction data	56,000	37,000
Generic	Attempting to cause a program or network suspended by feeding it the randomly generated data	40,000	18,871
Exploits	It contains different attacks of port scan, spam, and html files penetrations	33,393	11,132
Fuzzers	A technique in which a system security mechanism is bypassed stealthily to access a computer or its data	18,184	6,062
DoS	A malicious attempt to make a server or a network resource unavailable to users, usually by temporarily interrupting or suspending the services of a host connected to the Internet	12,264	4,089
Reconnaissance	The attacker knows of a security problem within an operating system or a piece of software and leverages that knowledge by exploiting the vulnerability	10,491	3,496

(Continued)

Table 6.1 (*Continued*) Description and Number of Records for Each Attack Category

Attack Category	Description	Number Records in Training Dataset	Number Records in Testing Dataset
Analysis	A technique works against all block-ciphers (with a given block and key size), without consideration about the structure of the block-cipher	2,000	677
Backdoor	Contains all strikes that can simulate attacks that gather information	1,746	583
Shellcode	A small piece of code used as the payload in the exploitation of software vulnerability	1,133	378
Worms	Attacker replicates itself in order to spread to other computers. Often, it uses a computer network to spread itself, relying on security failures on the target computer to access it	130	44
	Total	**175,341**	**82,332**

6.4.1.2 Dataset Features

There are 43 features in both the training and testing datasets, which are tabulated in Table 6.2. The description for each feature can be found at the original source ("The UNSW-NB15 Dataset Description," 2018).

6.4.2 Data Pre-Processing

6.4.2.1 Encoding for Categorical Types

As shown in Table 6.2, "proto," "service," and "state" features are categorical types, which need to be converted to numerical representations since the RF and DNN algorithms can only work with numerical features. The "proto" feature has 133 unique categories; however, the top 6 categories represent 92.34% of the records in the training dataset. For this particular feature, we will represent all categories

Table 6.2   Features of the UNSW-NB15 Dataset

	Feature Name	Type
1	id	Integer
2	dur	Float
3	proto	Categorical
4	service	Categorical
5	state	Categorical
6	spkts	Integer
7	dpkts	Integer
8	sbytes	Integer
9	dbytes	Integer
10	rate	Float
11	sttl	Integer
12	dttl	Integer
13	sload	Float
14	dload	Float
15	sloss	Integer
16	dloss	Integer
17	sinpkt	Float
18	dinpkt	Float
19	sjit	Float
20	djit	Float
21	swin	Integer
22	stcpb	Integer
23	dtcpb	Integer
24	dwin	Integer
25	tcprtt	Float

(Continued)

Table 6.2 (*Continued*) Features of the UNSW-NB15 Dataset

	Feature Name	Type
26	synack	Float
27	ackdat	Float
28	smean	Integer
29	dmean	Integer
30	trans_depth	Integer
31	response_body_len	Integer
32	ct_srv_src	Integer
33	ct_state_ttl	Integer
34	ct_dst_ltm	Integer
35	ct_src_dport_ltm	Integer
36	ct_dst_sport_ltm	Integer
37	ct_dst_src_ltm	Integer
38	is_ftp_login	Binary
39	ct_ftp_cmd	Integer
40	ct_flw_http_mthd	Integer
41	ct_src_ltm	Integer
42	ct_srv_dst	Integer
43	is_sm_ips_ports	Binary

outside of the top 6 as one numerical value, while the top 6 categories will have their own representation.

The "service" feature has 13 unique categories. In the training dataset, the records with its service labeled as "-" (which denotes all other services which had not been labeled as a well-known service such as "dns," "http," "smtp,") represent 53.71% of the data. The top 7 labeled services represent 46.14% of the data. Our approach here is to relabel all services which were not one of the top 7 labeled services as "other," including those which were originally labeled as "-".

The "state" feature has nine unique categories. The top 4 categories represent 99.94% of the records in the training dataset. For this feature, we will represent all categories outside the top 4 as one numerical value, while the top 4 categories will have their own representation.

Another reason for the abovementioned strategy of having a category that "catches all" other values is that when the model is presented with a value that has not seen before in its training, it will still be able to generalize and make a reasonable prediction.

Figure 6.3 shows the proportion of records with various "proto" values after our data transformation, Figure 6.4 shows the proportion of records with various "service" values after our data transformation, and Figure 6.5 shows the proportion of records with various "state" values after our data transformation.

In this research, we encode the three categorical features in two ways. The first way is using one-hot encoding, where we introduce one new feature for each of the top categories identified above, and one new feature for all other categories. The record with a category value equal to the new feature will have a boolean value of 1, while all the other new features will have a boolean value of 0.

The second way of encoding is converting each category to a numeric value, where the top categorical value will be assigned a numeric value of 1, the second top value a numeric value of 2, and so on, until the "catch all" category, which will assume the last numeric value.

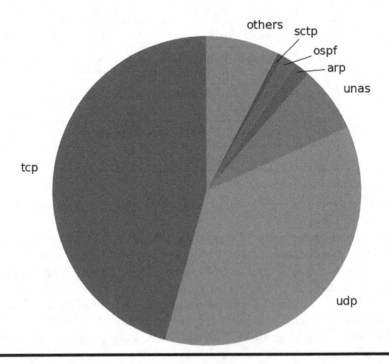

Figure 6.3   Number of records by "proto" values.

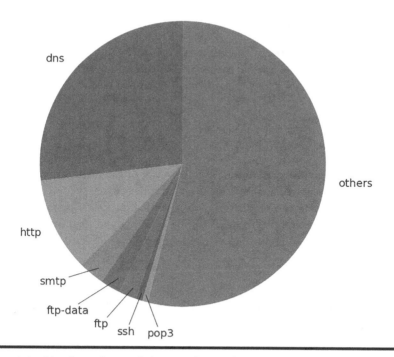

Figure 6.4   Number of records by "service" values.

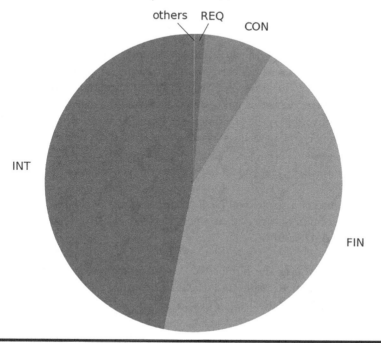

Figure 6.5   Number of records by "state" values.

6.4.2.2 Feature Scaling

In this research, features that are presented to the DNN model are scaled using the standardization method such that each scaled feature will have unit variance and a mean of zero. The formula for standardization is defined below:

$$x' = \frac{x - \overline{x}}{\sigma}$$

where x is an original feature value, $\overline{x}$ is the mean of the feature values, σ is the feature's standard deviation, and x' is the scaled value.

The reason why feature standardization is needed for DNN is that DNN uses gradient descent to minimize its losses during training, and gradient descent converges much better when all the features are in similar ranges ("Feature scaling," 2019).

6.4.2.3 Feature Dropping

The only feature that gets dropped in the research is the "id" column. This is just an identifier for each record, and it serves no purpose for our algorithms.

6.4.3 Performance Evaluation Metrics

Each prediction from the model falls into one of the following four outcomes:

1. True positive (TP)—the prediction of a record is attack and its actual label is attack; i.e., the model has correctly predicted the record as attack.
2. True negative (TN)—the prediction of a record is normal and its actual label is normal; i.e., the model has correctly predicted the record as normal.
3. False positive (FP)—the prediction of a record is attack, while its actual label is normal; i.e., the model has incorrectly predicted the record as attack.
4. False negative (FN)—the prediction of a record is normal, while its actual label is attack; i.e., the model has incorrectly predicted the record as normal.

The performance of a model will be evaluated in terms of the following metrics, which are used commonly in evaluating NIDSs (Aziz, Hanafi, & Hassanien, 2017; Yin, Zhu, Fei, & He, 2017):

1. Accuracy—this is the number of correct predictions made by the model over all predictions made, and it tells us the proportion of predictions made correctly. Mathematically, it is defined as

$$\text{Accuracy} = \frac{\text{TPs} + \text{TNs}}{\text{TPs} + \text{TNs} + \text{FPs} + \text{FNs}}$$

2. Precision—this is the number of correct attack predictions made by the model over all attack predictions made. Mathematically, it is defined as

$$Precision = \frac{TPs}{TPs + FPs}$$

3. Recall—this is the number of correct attack predictions made by the model over all actual attack records; it is the probability of attack detection. Recall is also known as the true positive rate (TPR), which is used for plotting the receiver-operating characteristic (ROC) curve. Mathematically, it is defined as

$$Recall = \frac{TPs}{TPs + FNs}$$

4. F-measure—this is the harmonic mean of precision and recall, which is a fair way to measure how good the combined precision and recall scores are. Mathematically, it is defined as

$$F\text{-measure} = \frac{2PR}{P + R}$$

5. Fall-out—this is the number of incorrect attack predictions made by the model over all actual normal records; it is the probability of false alarm. Fall-out is also known as the false positive rate (FPR), which is used for plotting the ROC curve. Mathematically, it is defined as

$$Fall\text{-out} = \frac{FPs}{FPs + TNs}$$

6.4.4 Implementation Setup

The ML and DNN models used in this research were provided by Scikit-learn ("scikit-learn Machine Learning in Python," n.d.) and Keras ("Keras: The Python Deep Learning Library," n.d.) libraries, both of which are Python libraries. Under the hood, Keras is using Google's TensorFlow ("TensorFlow: An end-to-end open source machine learning platform," n.d.) as its backend. Our code was written in Jupyter ("Project Jupyter," 2019) notebook, while results from the research were visualized using Matplotlib ("Matplotlib: Python plotting," 2019).

The versions of the software and libraries used in this research are summarized in Table 6.3.

Table 6.3   Versions of Software and Libraries Used

Software/Library	Version
Python	3.6.7
Jupyter	0.35.4
Numpy	1.13.3
Pandas	0.23.4
Matplotlib	2.2.3
Scikit-learn	0.20.2
TensorFlow	1.12.0
Keras	2.2.4

6.5 Results

In this section, the results from our investigations are presented and analyzed. In every experiment we conducted, we consistently used both RF and DNN algorithms to produce results that were comparable. Moreover, ROC plots are presented where appropriate to enhance the comparison via visual means.

The first experiment was to investigate whether one-hot encoding gives better results in comparison with numeric encoding. In this preliminary experiment, we created standard RF and DNN models without any optimizations applied to them. Only the column "label" in the datasets was used as the target for this investigation, and the column "attack_cat" was ignored.

Next, we tested out the one-attack-type-versus-normal approach, where each attack type in the "attack_cat" column has its own individual model. The idea behind this is to investigate how each model performs when given records are labeled either with its attack type or with its normal type. Additionally, we built an ensemble from these individual models and presented its results.

Following the one-attack-type-versus-normal approach, we tested out a slightly different approach, the one-attack-type-versus-all approach. In the one-attack-type-versus-all approach, each attack type also has its own individual model, but these models are trained and evaluated the given records that are labeled with any attack type. Similarly, we also built an ensemble model from these individual models and presented its results.

Lastly, we next investigated how we could get better performances out of RF and DNN models, by adjusting hyperparameters. For our optimized RF model, we also ranked the features according to their relative importance in terms of their influence on the prediction result.

6.5.1 One-Hot Encoding versus Numeric Encoding Using RF Classifier

In this first experiment, we created a default RF classifier model using Scikit-learn's RandomForestClassifier without supplying any custom parameters, except for random _ state which was set to a constant value for reproducibility. Our model is defined below:

```
RandomForestClassifier(bootstrap=True, class_weight=None,
   criterion='gini',
   max_depth=None, max_features='auto', max_leaf_nodes=None,
   min_impurity_decrease=0.0, min_impurity_split=None,
   min_samples_leaf=1, min_samples_split=2,
   min_weight_fraction_leaf=0.0, n_estimators=10, n_jobs=-1,
   oob_score=False, random_state=2019, verbose=0,
   warm_start=False)
```

("Forests of randomized trees," 2019)

The model was first trained using the one-hot encoded dataset, then trained using the numeric encoded dataset. Each of these models was evaluated against both the training and testing datasets; the results are tabulated in Table 6.4.

The fall-out scores (also known as FPR) are plotted against the accuracy scores (also known as TPR) in the ROC space shown in Figure 6.6.

The results suggest that RF classifier performs slightly better with one-hot encoded data over numeric encoded data.

Table 6.4   RF Classifier Results Trained and Evaluated Using One-Hot Encoded and Numeric Encoded Data

Encoding	Dataset	Accuracy (%)	Precision (%)	Recall (%)	F-Measure (%)	Fall-Out (%)
One-hot encoded	Training	99.6555	99.7328	99.7612	99.7470	0.5696
	Testing	87.6294	83.2183	97.1168	89.6320	23.9946
Numeric encoded	Training	99.6378	99.7302	99.7377	99.7340	0.5750
	Testing	87.8553	83.4245	97.2690	89.8164	23.6784

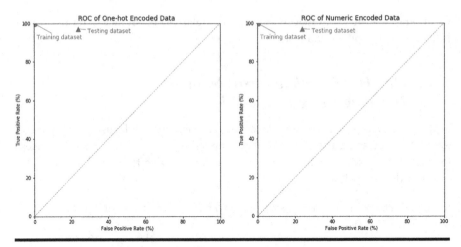

Figure 6.6 ROC of RF classifiers: one-hot encoded data versus numeric encoded data.

6.5.2 One-Hot Encoding versus Numeric Encoding Using DNN

In this first experiment with DNN, we created a model using Keras with 64 nodes in the input layer, 64 nodes in the hidden layer, and one node in the output layer for our classification problem. The number 64 was chosen because we have close to 64 features in our dataset. Our model is defined below:

```
model = Sequential()
model.add(Dense(64,input_dim=X_trn.shape[1],
  activation='relu'))
model.add(Dense(64, activation='relu'))
model.add(Dense(1, activation='sigmoid'))
model.compile(loss='binary_crossentropy', optimizer='adam',
  metrics=['accuracy'])
```

(Brownlee, 2016; "Getting started with the Keras Sequential model," n.d.)

Training was first performed using the one-hot encoded data, then using the numeric encoded dataset, both with 300 epochs and with batch size of 4,000. These two models were evaluated against both the training and testing datasets; the results are tabulated in Table 6.5.

The fall-out scores are plotted against the accuracy scores in the ROC space shown in Figure 6.7.

The results suggest that the DNN model gives almost identical results regardless of which data encoding was used, with numeric encoded dataset giving a slight, and probably negligible, advantage.

Table 6.5   DNN Results Trained and Evaluated Using One-Hot Encoded and Numeric Encoded Data

Encoding	Dataset	Accuracy (%)	Precision (%)	Recall (%)	F-Measure (%)	Fall-Out (%)
One-hot encoded	Training	95.3605	96.1581	97.0614	96.6076	8.2643
	Testing	86.7378	82.2020	96.8918	88.9445	25.7027
Numeric encoded	Training	95.2966	95.2544	97.9705	96.5933	10.4018
	Testing	85.4844	80.2140	97.7477	88.1171	29.5405

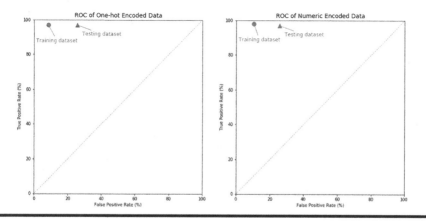

Figure 6.7   ROC of DNN classifiers: one-hot encoded data versus numeric encoded data.

6.5.3 One-Attack-Type-versus-Normal RF Classifiers

The previous investigation discarded the type of attack; it only looked at whether a record is a normal traffic or an attack traffic. Next in this investigation, the one-attack-type-versus-normal approach takes into account the attack type for each record. Building up a model for each attack category and analyzing its performance will give us insight into the difficulty in predicting an attack type compared to others.

In this investigation, we trained one model for each of the nine attack types. Each of these models was trained from a subset of the full training dataset; this subset included only records where attack is the type we were training the model for, plus records which are normal traffic. Similarly, each model was also tested against a subset of the full testing dataset, which consisted only of records of that attack type and normal traffic.

We used RF classifier with the same hyperparameters as that which were used in the first investigation. One-hot encoded training and testing datasets were chosen for this experiment. The performance metrics from each of the nine trained RF models against the training dataset are tabulated in Table 6.6, while the performance metrics against the testing dataset are tabulated in Table 6.7.

The ROC plot for the individual models using the training and the testing datasets is shown in Figures 6.8 and 6.9, respectively.

Table 6.6 RF Performance Metrics for One-versus-Normal Models Using Training Dataset

Attack Type	Accuracy (%)	Precision (%)	Recall (%)	F-Measure (%)	Fall-Out (%)
Backdoor	**99.9948**	**100.0000**	99.8282	99.9140	0.0000
Analysis	99.9638	99.9495	99.0000	99.4725	0.0018
Fuzzers	99.2303	98.4646	98.3942	98.4294	0.4982
Shellcode	99.9720	99.7329	98.8526	99.2908	0.0054
Reconnaissance	99.9850	99.9619	99.9428	99.9523	0.0071
Exploits	99.9396	99.9700	99.8682	99.9191	0.0179
DoS	99.9517	99.9510	99.7798	99.8653	0.0107
Worms	99.9911	**100.0000**	96.1538	98.0392	**0.0000**
Generic	99.9760	99.9975	**99.9450**	**99.9712**	0.0018

Best results highlighted in bold.

Table 6.7 RF Performance Metrics for One-versus-Normal Models Using Testing Dataset

Attack Type	Accuracy (%)	Precision (%)	Recall (%)	F-Measure (%)	Fall-Out (%)
Backdoor	99.8882	95.7699	97.0840	96.4225	0.0676
Analysis	98.5455	55.8161	91.4328	69.3169	1.3243
Fuzzers	79.8662	38.6133	72.9462	50.4967	19.0000
Shellcode	99.1385	54.4444	90.7407	68.0556	0.7757
Reconnaissance	99.4641	98.2632	95.4805	96.8519	0.1595
Exploits	97.3032	91.4657	97.4308	94.3541	2.7351
DoS	98.6420	90.9344	95.9159	93.3587	1.0568
Worms	**99.9541**	88.5714	70.4545	78.4810	**0.0108**
Generic	99.5579	**99.7330**	**98.9561**	**99.3430**	0.1351

Best results highlighted in bold.

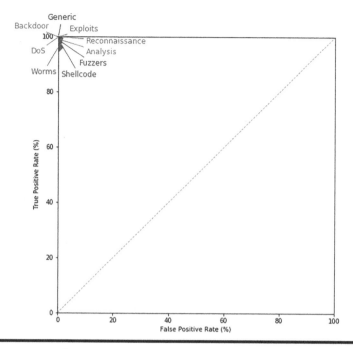

Figure 6.8   ROC of RF one-versus-normal models using training dataset.

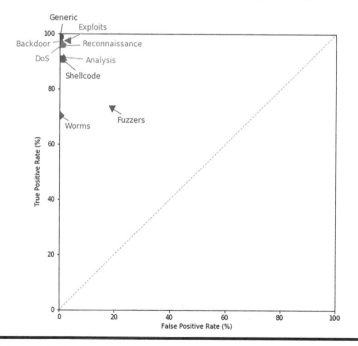

Figure 6.9   ROC of RF one-versus-normal models using testing dataset.

With the nine individual models that had been trained to recognize their attack category, we built an ensemble model which works as follows. Given a record, it makes a prediction by consulting each of these models. If one of the individual models predicts the record as an attack, then this model will predict it as an attack. Conversely, if none of the individual models predicts the record as an attack, then this model will predict it as a normal record. This ensemble model was being evaluated against the training and testing datasets; the results are tabulated in Table 6.8. The ROC plot of these metrics is shown in Figure 6.10.

Table 6.8   RF Performance Metrics for Ensemble of One-versus-Normal Models

Dataset	Accuracy (%)	Precision (%)	Recall (%)	F-Measure (%)	Fall-Out (%)
Training	99.6361	99.7477	99.7176	99.7327	0.5375
Testing	88.3120	84.3323	96.7462	90.1137	22.0216

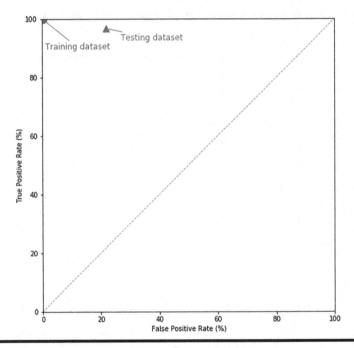

Figure 6.10   ROC of RF ensemble of one-versus-normal models.

6.5.4 One-Attack-Type-versus-Normal DNN Classifiers

We extended the same one-attack-type-versus-normal approach to DNN. As before, the training and testing datasets were prepared the same way. The performance metrics from each of the nine trained DNN models against the training dataset are tabulated in Table 6.9, while the performance metrics against the testing dataset are tabulated in Table 6.10.

The ROC plot for the individual models using the training and the testing datasets is shown in Figures 6.11 and 6.12, respectively.

As before, we went ahead with building an ensemble model which consults the individual trained models for each type of the attack. This ensemble model was being evaluated against the training and testing datasets, whose results are tabulated in Table 6.11. The ROC plot of these metrics is shown in Figure 6.13.

6.5.5 One-Attack-Type-versus-All RF Classifiers

The one-attack-type-versus-all approach is similar to the one-attack-type-versus-normal approach, with the exception that each individual model is trained to distinguish its attack type from among all types of records, not just the normal records. As such, these models are trained using the full dataset, whereby labels are modified such that records of the attack type being trained for are marked 1,

Table 6.9 DNN Performance Metrics for One-versus-Normal Models Using Training Dataset

Attack Type	Accuracy (%)	Precision (%)	Recall (%)	F-Measure (%)	Fall-Out (%)
Backdoor	99.9740	99.4289	99.7136	99.5711	0.0179
Analysis	99.9069	98.5045	98.8000	98.6520	0.0536
Fuzzers	91.7853	82.0044	85.1793	83.5617	6.0696
Shellcode	99.6744	93.3211	90.0265	91.6442	0.1304
Reconnaissance	99.8030	99.2208	99.5329	99.3766	0.1464
Exploits	98.9954	99.2274	98.0744	98.6475	0.4554
DoS	99.5928	99.6932	98.0349	98.8571	0.0661
Worms	**99.9786**	96.8254	93.8462	95.3125	**0.0071**
Generic	99.9542	**99.9650**	**99.9250**	**99.9450**	0.0250

Best results highlighted in bold.

Table 6.10 DNN Performance Metrics for One-versus-Normal Models Using Testing Dataset

Attack Type	Accuracy (%)	Precision (%)	Recall (%)	F-Measure (%)	Fall-Out (%)
Backdoor	99.1911	71.1684	80.4460	75.5233	0.5135
Analysis	97.4759	40.9151	91.1374	56.4760	2.4081
Fuzzers	77.4418	35.1593	71.3626	47.1088	21.5622
Shellcode	98.5366	39.6064	85.1852	54.0722	1.3270
Reconnaissance	97.8467	82.3631	95.5092	88.4503	1.9324
Exploits	95.4396	86.5701	95.0234	90.6000	4.4351
DoS	98.0627	87.3610	94.1551	90.6309	1.5054
Worms	**99.9190**	63.4615	75.0000	68.7500	**0.0514**
Generic	99.1892	**98.6991**	**98.9031**	**98.8010**	0.6649

Best results highlighted in bold.

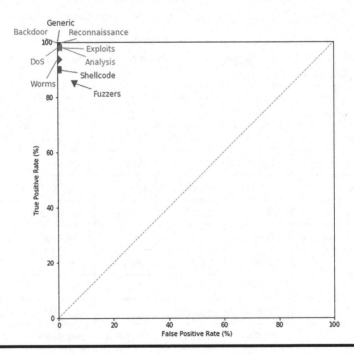

Figure 6.11 ROC of DNN one-versus-normal models using training dataset.

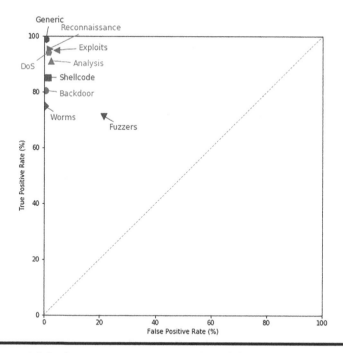

Figure 6.12 ROC of DNN one-versus-normal models using testing dataset.

Table 6.11 DNN Performance Metrics for Ensemble of One-versus-Normal Models

Dataset	Accuracy (%)	Precision (%)	Recall (%)	F-Measure (%)	Fall-Out (%)
Training	96.1886	96.9581	97.4577	97.2072	6.5161
Testing	86.1998	81.9111	96.1749	88.4718	26.0216

while all other records, including those of other attack types, are marked 0. The performance metrics from each of the nine trained RF models against the training dataset are tabulated in Table 6.12, while the performance metrics against the testing dataset are tabulated in Table 6.13.

The ROC plot for the individual models using the training and the datasets is shown in Figures 6.14 and 6.15, respectively.

As per the one-attack-type-versus-normal investigation, an ensemble of one-attack-type-versus-all models was built and evaluated, and its results are tabulated in Table 6.14. The ROC plot of these metrics is shown in Figure 6.16.

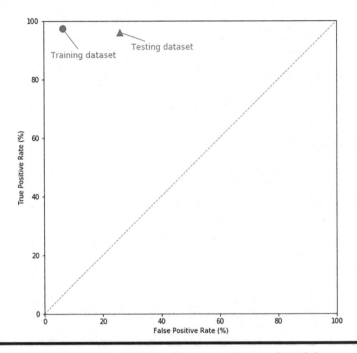

Figure 6.13 ROC of DNN ensemble of one-versus-normal models.

Table 6.12   RF Performance Metrics for One-versus-All Models Using Training Dataset

Attack Type	Accuracy (%)	Precision (%)	Recall (%)	F-Measure (%)	Fall-Out (%)
Backdoor	99.1998	89.7912	22.1649	35.5535	0.0253
Analysis	99.1565	92.9160	28.2000	43.2681	0.0248
Fuzzers	98.7299	98.2522	89.3423	93.5856	0.1839
Shellcode	99.9504	98.7873	93.4687	96.0544	0.0075
Reconnaissance	98.8810	99.4549	81.7463	89.7353	0.0285
Exploits	92.6093	84.8251	74.5246	79.3420	3.1364
DoS	94.2803	98.1059	18.5828	31.2470	0.0270
Worms	**99.9875**	93.5484	89.2308	91.3386	**0.0046**
Generic	99.7610	**99.9722**	**98.9800**	**99.4736**	0.0081

Best results highlighted in bold.

Table 6.13   RF Performance Metrics for One-versus-All Models Using Testing Dataset

Attack Type	Accuracy (%)	Precision (%)	Recall (%)	F-Measure (%)	Fall-Out (%)
Backdoor	96.7485	2.1481	8.0617	3.3923	2.6190
Analysis	98.0081	0.0000	0.0000	0.0000	1.1794
Fuzzers	88.2233	29.0620	41.6034	34.2198	8.0713
Shellcode	99.5239	48.0114	44.7090	46.3014	0.2233
Reconnaissance	98.8850	95.0384	77.8032	85.5615	0.1801
Exploits	92.5752	78.1935	62.5225	69.4853	2.7261
DoS	95.2449	77.3585	6.0161	11.1641	0.0920
Worms	**99.9502**	63.6364	15.9091	25.4545	**0.0049**
Generic	99.2166	**99.8796**	**96.6986**	**98.2634**	0.0347

Best results highlighted in bold.

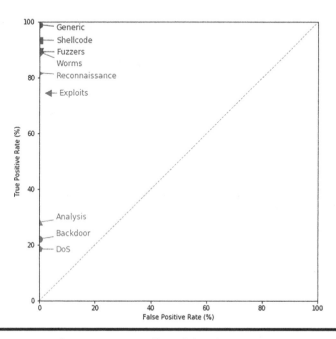

Figure 6.14   ROC of RF one-versus-all models using training dataset.

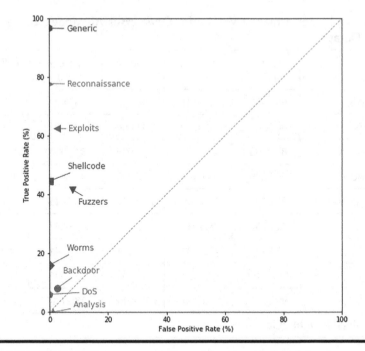

Figure 6.15 ROC of RF one-versus-all models using testing dataset.

Table 6.14 RF Performance Metrics for Ensemble of One-versus-All Models

Dataset	Accuracy (%)	Precision (%)	Recall (%)	F-Measure (%)	Fall-Out (%)
Training	87.8842	99.7253	82.4260	90.2542	0.4839
Testing	80.4717	84.5188	79.0038	81.6683	17.7297

6.5.6 One-Attack-Type-versus-All DNN Classifiers

The DNN version of the one-attack-type-versus-all approach was also being investigated. The performance metrics from each of the nine trained DNN models against the training dataset are tabulated in Table 6.15, while the performance metrics against the testing dataset are tabulated in Table 6.16.

The ROC plot for the individual models using the training and the testing datasets is shown in Figures 6.17 and 6.18, respectively.

Similarly, the DNN ensemble version of one-versus-all models was built and evaluated, and the results are tabulated in Table 6.17. The ROC plot of these metrics is shown in Figure 6.19.

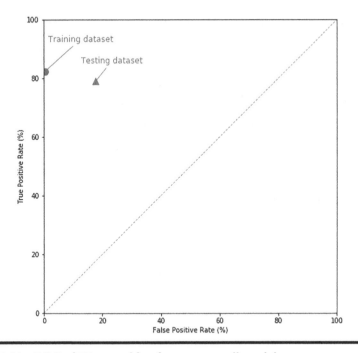

Figure 6.16 ROC of RF ensemble of one-versus-all models.

Table 6.15 DNN Performance Metrics for One-versus-All Models Using Training Dataset

Attack Type	Accuracy (%)	Precision (%)	Recall (%)	F-Measure (%)	Fall-Out (%)
Backdoor	99.1348	88.0399	15.1775	25.8915	0.0207
Analysis	99.1154	96.4803	23.3000	37.5352	0.0098
Fuzzers	95.0029	79.9187	69.2037	74.1762	2.0120
Shellcode	99.6013	67.1136	75.1103	70.8871	0.2394
Reconnaissance	98.2845	92.7844	77.3425	84.3627	0.3828
Exploits	90.1529	84.6326	59.0094	69.5356	2.5206
DoS	93.5400	75.9845	11.1709	19.4782	0.2655
Worms	**99.9721**	90.0990	70.0000	78.7879	**0.0057**
Generic	99.6749	**99.9013**	**98.6725**	**99.2831**	0.0288

Best results highlighted in bold.

Table 6.16 DNN Performance Metrics for One-versus-All Models Using Testing Dataset

Attack Type	Accuracy (%)	Precision (%)	Recall (%)	F-Measure (%)	Fall-Out (%)
Backdoor	98.2182	5.7114	9.7770	7.2106	1.1511
Analysis	97.8125	7.4887	14.6233	9.9050	1.4978
Fuzzers	87.2844	23.7585	32.9099	27.5953	8.3939
Shellcode	98.7162	20.0881	60.3175	30.1388	1.1067
Reconnaissance	98.5777	85.6705	79.8627	82.6647	0.5924
Exploits	91.6181	71.8431	62.5045	66.8492	3.8301
DoS	94.6121	34.5778	9.5133	14.9214	0.9407
Worms	**99.9429**	42.1053	18.1818	25.3968	**0.0134**
Generic	99.0818	**99.2336**	**96.7410**	**97.9715**	0.2222

Best results highlighted in bold.

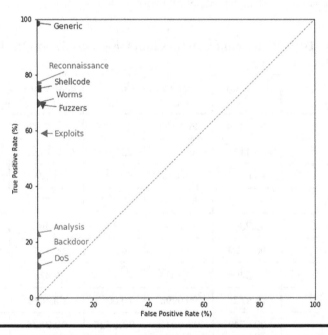

Figure 6.17 ROC of DNN one-versus-all models using training dataset.

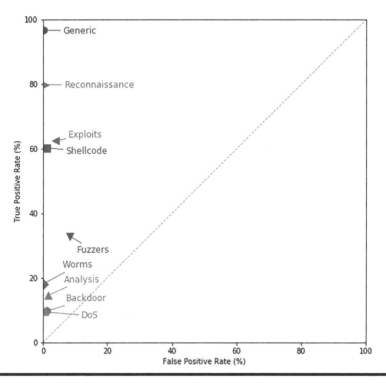

Figure 6.18   ROC of DNN one-versus-all models using testing dataset.

Table 6.17   DNN Performance Metrics for Ensemble of One-versus-All Models

Dataset	Accuracy (%)	Precision (%)	Recall (%)	F-Measure (%)	Fall-Out (%)
Training	79.5393	96.4071	72.6456	82.8564	5.7696
Testing	78.4361	82.0376	77.8898	79.9099	20.8946

6.5.7 Optimizing RF Classifier

Among all the models experimented, the default configured RF classifier using one-hot encoded data presented in the first investigation is shown to be promising. We looked at ways to optimize the RF classifier and see how much we can improve on this model. With some experimentations in tweaking hyperparameters, our findings showed us that adjusting n_estimators, max_depth, and max_features

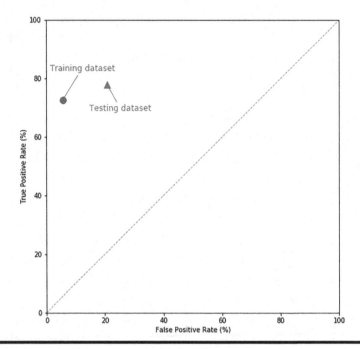

Figure 6.19 ROC of DNN ensemble of one-versus-all models.

gave us better results. We finally settled on n_estimators == 107, max_depth == 60, and max_features == 2, defined as:

```
RandomForestClassifier(bootstrap=True, class_weight=None,
    criterion='gini',
    max_depth=60, max_features=2, max_leaf_nodes=None,
    min_impurity_decrease=0.0, min_impurity_split=None,
    min_samples_leaf=1, min_samples_split=2,
    min_weight_fraction_leaf=0.0, n_estimators=6, n_jobs=-1,
    oob_score=False, random_state=2019, verbose=0,
    warm_start=False)
```

The results are tabulated in Table 6.18. The ROC plot of these metrics is shown in Figure 6.20.

Table 6.18 Optimized RF Classifier Results Using One-Hot Encoded Data

Dataset	Accuracy (%)	Precision (%)	Recall (%)	F-Measure (%)	Fall-Out (%)
Training	99.4034	99.6775	99.4453	99.5612	0.6857
Testing	88.0605	83.9901	96.7595	89.9237	22.5973

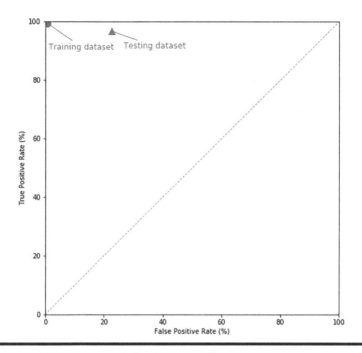

Figure 6.20   ROC of our optimized RF classifier.

6.5.8 Feature Importances of the Optimized RF Classifier

All feature importances of the above optimized RF classifier are plotted in the bar graph in Figure 6.21. The top 10 feature importances are tabulated in Table 6.19.

6.5.9 Optimizing DNN Classifier

For the purpose of comparing against the optimized RF classifier, we also optimized the DNN model. We tried varying the number of hidden layers and varying the number of nodes at each layer. One of the best results we got was using a network of three layers with 200 units at its first two layers. It was created using the following code:

```
model = Sequential()
model.add(Dense(200,input_dim=X_trn.shape[1],
    activation='relu'))
model.add(Dense(200, activation='relu'))
model.add(Dense(1, activation='sigmoid'))
model.compile(loss='binary_crossentropy', optimizer='adam',
    metrics=['accuracy'])
```

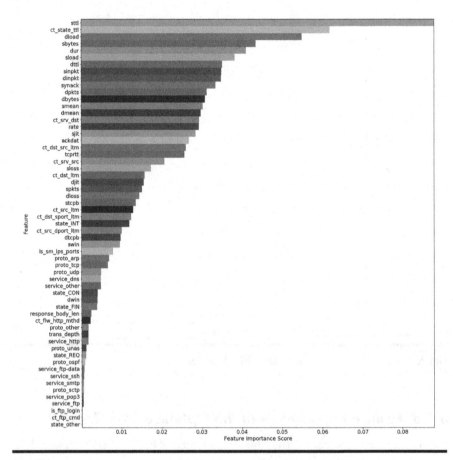

Figure 6.21 Feature importances of our optimized RF classifier.

Table 6.19 Top 10 Feature Importances of Our Optimized RF Classifier

Feature	Score (Mean Decrease in Gini)
sttl	0.087840
ct_state_ttl	0.061850
dload	0.054946
sbytes	0.043473
dur	0.041077
sload	0.038244

(*Continued*)

Table 6.19 (*Continued*)   Top 10 Feature Importances of Our Optimized RF Classifier

Feature	Score (Mean Decrease in Gini)
dttl	0.035161
sinpkt	0.034921
dinpkt	0.034853
synack	0.033477

The metrics were slightly better compared to the initial DNN model we built; these scores are tabulated in Table 6.20. The ROC plot of these metrics is shown in Figure 6.22.

Table 6.20   Optimized DNN Classifier Results Using One-Hot Encoded Data

Dataset	Accuracy (%)	Precision (%)	Recall (%)	F-Measure (%)	Fall-Out (%)
Training	95.0639	96.4201	96.3240	96.3720	7.6214
Testing	87.8310	83.4536	97.1632	89.7881	23.6027

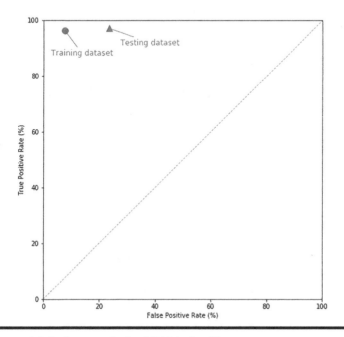

Figure 6.22   ROC of our optimized DNN classifier.

6.6 Discussions

The results from our experiments revealed a number of interesting observations regarding the datasets as well as the algorithms we had chosen for this research.

6.6.1 Data Encoding

When it comes to encoding categorical features, namely, "proto," "service," and "state" features, we found that one-hot encoding gave us slightly better results compared to numerical encoding in most of the metrics we used. This was observed in both RF and DNN models.

6.6.2 One-Attack-Type-versus-Normal Approach

The individual models we built using RF classifiers gave us very decent results against the training dataset, but did not perform too well against the testing dataset. The models that performed particularly badly were the ones trained for Fuzzers and Worms attack types. The ensemble model built from these individual RF classifiers did very well against the training set, but did not perform well against the testing set. These discrepancies highlight an overfitting issue with this approach, implying that the model is unable to give an accurate prediction for cases it has not seen before.

Similarly, the individual models we built using DNN performed much worse against the testing dataset than against the training dataset. The Fuzzers and Worms models were again two of the worst ones. The ensemble model built from these individual DNN models also did not perform well against the testing dataset. We can conclude that our ensemble DNN model constructed with this approach has a similar overfitting problem.

From the above experiments, we learned that the one-attack-type-versus-normal approach not only adds complexity to the architecture, but also does not give us a reliable prediction model. An interesting observation we got out of this exercise was that it is difficult to get accurate prediction for Fuzzers and Worms attack types.

6.6.3 One-Attack-Type-versus-All Approach

In the one-attack-type-versus-all approach, the individual models we built using RF classifiers performed quite badly with their recall rates, particularly against the testing dataset. The ensemble version also suffered from a high fall-out rate against the testing dataset.

The individual DNN models built using this approach performed similarly to their RF counterparts; they also suffer from high recall rates. The ensemble model built from these individual DNN models performed worse than its RF counterpart.

Against the testing dataset, all the five metrics used in this research were of an unacceptable level.

From this, we learned that the one-attack-type-versus-all approach is unsuitable for the purpose of this research. While most of the individual DNN models performed rather poorly, the only one that did well was the model built for Generic attack type. This exercise has indicated to us that it is relatively easy to get an accurate prediction for Generic attack type.

6.6.4 Feature Importances

According to our RF model, the 10 most important features are "sttl," "ct_state_ttl," "dload," "sbytes," "dur," "sload," "dttl," "sinpkt," "dinpkt," and "synack." These features give a combined weighting of 0.465842, close to 50%. The 30 least important features combine to give a weighting of 0.107293. Interestingly, all categorical encoded features fall into this group. These are the features that can be dropped without a detrimental effect on the prediction outcome.

6.6.5 RF versus DNN

From the experiments we had conducted, there is very little to separate these two algorithms in terms of their prediction results, although RF is giving slightly better results in all of the metrics except the recall rate. This is tabulated in Table 6.21. Note that both models are producing very high recall rates of over 96%, which is important in our network intrusion detection problem. In other words, they are able to detect over 96% of the attack records correctly. However, the fall-out rates for both models are not great, raising false alarms around 22%–23% of the time. Note that the fall-out rates can be improved by adjusting the classification threshold, but this will come at a cost of a poorer recall rate.

Table 6.21 Comparison of RF and DNN Classifiers against Testing Dataset

Model	Accuracy (%)	Precision (%)	Recall (%)	F-Measure (%)	Fall-Out (%)
RF	**88.0605**	**83.9901**	96.7595	**89.9237**	**22.5973**
DNN	87.8310	83.4536	**97.1632**	89.7881	23.6027
Difference	0.2295	0.5365	0.4037	0.1356	1.0054

Best results highlighted in bold.

6.7 Conclusions

Today, networks are under unprecedented threats. Traditional rule-based NIDSs are unable to detect attacks that are constantly evolving. This research has investigated the effectiveness of applying ML-based and deep learning-based techniques to network intrusion detection.

RF and DNN algorithms were used in the study; they were compared and contrasted for their effectiveness against the UNSW-NB15 dataset. One-hot encoding and numerical encoding were used to encode categorical values; these two types of encoding were compared for their effectiveness.

Additional approaches, such as one-attack-type-versus-normal and one-attack-type-versus-all approaches, were investigated. Although these approaches did not give us better predictive models, they were able to, however, highlight to us which attack types were particularly hard to predict accurately.

From the RF model, we were able to discover which were the top features in terms of their importance. We were also able to identify which features contribute very little to the prediction algorithm.

As a result of our research, we can conclude that ML-based and deep learning-based techniques are effective in detecting network intrusions, both giving us satisfactory accuracy rates of over 87%.

References

Aziz, A., Hanafi, S., & Hassanien, A. (2017). Comparison of classification techniques applied for network intrusion detection and classification. *Journal of Applied Logic: Part A*, 24, 109–118. doi:10.1016/j.jal.2016.11.018.

Brownlee, J. (2016, Jun 2). Multi-Class Classification Tutorial with the Keras Deep Learning Library. Retrieved from https://machinelearningmastery.com/multi-class-classification-tutorial-keras-deep-learning-library/.

Deep learning. (2019, May 10). In Wikipedia. Retrieved from https://en.wikipedia.org/wiki/Deep_learning.

Dutt, D. (2018, Jan 10). 2018: The Year of the AI-Powered Cyberattack. Retrieved from www.csoonline.com/article/3246196/cyberwarfare/2018-the-year-of-the-ai-powered-cyberattack.html.

Elazari, K. (2017, Dec 28). Hackers Are on the Brink of Launching a Wave of AI Attacks. Retrieved from www.wired.co.uk/article/hackers-ai-cyberattack-offensive.

Forests of randomized trees. (2019). Retrieved from https://scikit-learn.org/stable/modules/ensemble.html#forest.

Ensemble learning. (2019, May 10). In *Wikipedia*. Retrieved from https://en.wikipedia.org/wiki/Ensemble_learning.

Feature scaling. (2019, May 30). In *Wikipedia*. Retrieved from https://en.wikipedia.org/wiki/Feature_scaling.

Getting started with the Keras Sequential model. (2019, June 06). Retrieved from https://keras.io/getting-started/sequential-model-guide/.

Keras: The Python Deep Learning Library. (2019, June 06). Retrieved from https://keras.io.

KDD Cup 1999 Data. (1999, Oct 28). Retrieved from http://kdd.ics.uci.edu/databases/kddcup99/kddcup99.html.

Matplotlib: Python plotting. (2019, Apr 28). Retrieved from https://matplotlib.org.

Moustafa, N. & Slay, J. (2015a). 4th International Workshop on Building Analysis Datasets and Gathering Experience Returns for Security (BADGERS) Kyoto, Japan 2015 Nov. 5. In *The Significant Features of the UNSW-NB15 and the KDD99 Data Sets for Network Intrusion Detection Systems* (pp. 25–31). IEEE. doi:10.1109/BADGERS.2015.014.

Moustafa, N. & Slay, J. (2015b). Military Communications and Information Systems Conference (MilCIS) Canberra, Australia 2015 Nov. 10–2015 Nov. 12. In *UNSW-NB15: A Comprehensive Data Set for Network Intrusion Detection Systems (UNSW-NB15 Network Data Set)* (pp. 1–6). IEEE. doi:10.1109/MilCIS.2015.7348942.

Moustafa, N. & Slay, J. (2016). The evaluation of network anomaly detection systems: Statistical analysis of the unsw-nb15 data set and the comparison with the kdd99 data set. *Information Security Journal: A Global Perspective*, 25(1–3), 18–31. doi:10.1080/19393555.2015.1125974.

Niyaz, Q., Sun, W., Javaid, A. Y., & Alam, M. (2016). A Deep Learning Approach for Network Intrusion Detection System. *EAI International Conference on Bio-Inspired Information and Communications Technologies (BICT)* (2015). doi:10.4108/eai.3-12-2015.2262516.

NSL-KDD dataset. (2019, June 06). Retrieved from www.unb.ca/cic/datasets/nsl.html.

Project Jupyter. (2019, Apr 12). Retrieved from https://jupyter.org.

scikit-learn Machine Learning in Python. (2019, June 06). Retrieved from https://scikit-learn.org/stable/.

Sobers, R. (2019, Apr 17). 60 Must-Know Cybersecurity Statistics for 2019. Retrieved from www.varonis.com/blog/cybersecurity-statistics/.

Tang, T. A., Mhamdi, L., McLernon, D., Zaidi, S. A. R., & Ghogho, M. (2016). Deep Learning Approach for Network Intrusion Detection in Software Defined Networking. *2016 International Conference on Wireless Networks and Mobile Communications (Wincom)* (pp. 258–263). IEEE. doi:10.1109/WINCOM.2016.7777224.

The UNSW-NB15 Dataset Description. (2018, Nov 14). Retrieved from www.unsw.adfa.edu.au/unsw-canberra-cyber/cybersecurity/ADFA-NB15-Datasets/.

TensorFlow: An end-to-end open source machine learning platform. (2019, June 06). Retrieved from www.tensorflow.org.

Yin, C., Zhu, Y., Fei, J., & He, X. (2017). A deep learning approach for intrusion detection using recurrent neural networks. *IEEE Access*, 5. doi:10.1109/ACCESS.2017.2762418.

Chapter 7

The Kernel-Based Online Anomaly Detection Algorithm: Detailed Derivation and Development

Salva Daneshgadeh
Middle East Technical University

Tarem Ahmed
Independent University Bangladesh (IUB)

Al-Sakib Khan Pathan
Southeast University, Dhaka, Bangladesh

Contents

7.1 Introduction

Nowadays, network and communication technologies are inextricably tied to our everyday personal and professional lives. Remote access technologies are widely used in an ever-expanding diversity of fields encompassing education, health-care, automobiles, entertainment, energy, industrial products, smart offices, and smart homes. The popularity of these technologies encourages all players in the sector to develop new connected and embedded objects which introduced the concept of Internet of Things (IoT) [1]. According to a report by market research company Statista [2], industrial IoT adoption has reached 45% in America, 33% in EMEA (Europe, the Middle East, and Africa), and 22% in the Asia Pacific. The ultimate objective of IoT is to create a world of smart and connected objects which would communicate with each other and with human beings to enrich and make their interactions easier. All these modern concepts depict amazing oppor-tunities for scientists and users alike. Unfortunately, whenever new opportuni-ties emerge, new vulnerabilities also concurrently emerge. Therefore, all societies which are run based on the technology, from the perspective of government, business, industry, academia, or a home user, are prone to greater threats along with the greater benefits. For example, connected systems and IoT devices can be compromised and forced to operate in the manner in which they are not deigned to. Specially, typical IoT devices with limited storage, capacity, and computation power are more vulnerable to attacks rather than other Internet-enabled devices. As an example, cameras installed on smart televisions may be manipulated and remotely redirected for spying elsewhere. Various types of small smart devices in different geographical parts of the globe may be captured by a rogue *master bot* to control and command these *zombies* to perform a distributed Denial of Service (DDoS) attack.

Therefore, it is crucial to develop security mechanisms to protect information and technological infrastructures. There have been many such methods developed by both academia and industry [3–9].

In many works in existing literatures, the terms *cyber security* and *information security* are often used interchangeably. In general, information security concen-trates on protecting information assets, while cyber security encompasses more dimensions, including human resources and ethics [10]. The aim of information security is to protect the triad of confidentiality, integrity, and availability (CIA) of data [11]. Availability means that information should be available whenever it is needed by authorized people. Integrity means that information should only be altered by authorized people or otherwise, would be exactly the same as was pro-duced or sent by a sender. Confidentiality means that information should only be accessible by authorized people. Cyber security provides authenticity and non-repudiation in addition to the CIA triad of information security. Authenticity ensures the identity of the source. Non-repudiation means that the party of the communication cannot deny the authenticity of her/his signature. Moreover, cyber security can protect the privacy of the users.

We briefly summarize here the intersection of cyber security and information security as information and communication technology security.

7.1.1 Terminologies

Assets, threat, vulnerability, risk, and attacks are fundamental concepts which should be understood before studying a writing on cyber security. The formal definitions of these terms are as follows [12]:

- **Asset:** An equipment or human resource of the organization.
- **Threat:** A shared term for people and objects which can expose a potential danger to assets.
- **Attack:** A deliberate act that exploits vulnerability.
- **Vulnerability:** A weakness or fault which can be exploited by a threat actor.
- **Risk:** A probability of happening an unwanted/undesired event multiplying the adverse expected damage of the event to the organization.

Figure 7.1 presents an infographic demonstrating the relationship between threat, vulnerability, risk, and asset. Threat exploits vulnerability which leads to risk, which in turn damages assets.

Ref [13] classified attacks into 11 different groups: Virus, Worm, Trojan, DoS (Denial of Service), Network Attack, Password Attack, Physical Attack, Information Gathering, User to Root (U2R), Remote to Local (R2L), and Probe Attack.

Correspondingly, they classified different methods of detecting these attacks.

7.1.2 Anomaly Detection Methods

Cyber attack detection systems are either placed in a single workstation, creating a host-based intrusion detection system (HIDS), or placed as stand-alone devices on a network, to form a network-based intrusion detection system (NIDS). Both HIDSs and NIDSs are classified into two groups called signature/misused-based IDSs and anomaly-based IDSs. Signature-based attack detection methods compare the incoming network traffic with the patterns of known attacks in their database and an alarm is raised if a match is found. The performance of signature-based

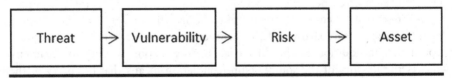

Figure 7.1 The relationship between threat, vulnerability, risk, and asset.

attack detection methods is close to 100% in detecting known attacks, but they cannot detect zero-day attacks [14]. Snort is a well-known signature-based (rule-based) NIDS, and Tripwire is an example of a signature-based HIDS [15].

Anomaly detection methods are employed to reveal behaviors, which constitute short-term deviations from the normal pattern of the network. Anomalies are often broadly classified into three types [3,15]:

- **Point Anomaly**: When a specific data sample differs from the normal pattern of data samples.
- **Contextual Anomaly**: When a data sample differs from the normal pattern of data sample under a specific condition or situation, for example, the number of GET requests to the website of a university during a semester registration period.
- **Collective Anomaly**: When a group of similar samples act differently from the data sample, for example, a DDoS attack.

Subsequently, NIDSs are classified into six groups [13]:

- Statistical,
- Classification-based,
- Clustering-based.
- Soft computing (e.g., genetic algorithm [GA]-based, artificial neural network [ANN]-based, Fuzzy set, Rough set, Ant Colony, Automatic Identification System [AIS]),
- Knowledge-based (e.g., Rule and Expert Systems, Ontology, and Logic-based),
- Combination learning (e.g., Assembly-based, Fusion-based, Hybrid).

7.1.3 Machine Learning and Anomaly Detection

Recent cyber security technologies embrace machine learning approaches to identify abnormal activities in network traffic statistics. However, many researchers believe that machine learning will never be a silver bullet for cyber security because attackers always manage to find a way to fool and evade detection engines. We believe that the speed of the algorithm to detect any alteration in network traffic also reflects its efficiency level. These requirements emphasize the advantage of online anomaly detection methods over traditional block-based schemes.

In general, classification and clustering techniques are popular machine learning approaches for detecting network anomalies. Classification methods are supervised which require labeled data for training. Regression, support vector machines, and random trees are examples of supervised machine learning algorithms. On the other hand, clustering methods are unsupervised which do not require labeled data. k-means, hierarchical clustering, and kth nearest neighbor are the examples of unsupervised machine learning algorithms.

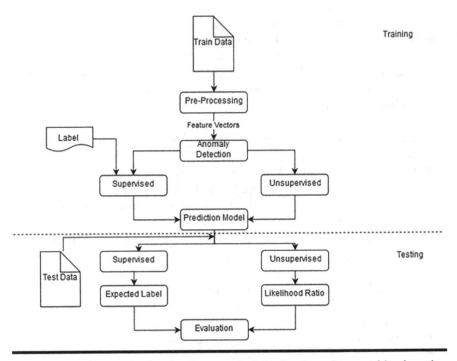

Figure 7.2 Framework of network anomaly detection based on machine learning approaches.

Figure 7.2 demonstrates the high-level view of network anomaly detection using supervised and unsupervised machine learning approaches.

All the algorithms mentioned above work in offline modes, which means that they cannot update their data repository in real time. Direct extensions and extrapolations of traditional offline methods for real-time use have also proved ineffective in the past [16,17]. Therefore, if the scope of normal behavior changes during runtime, the algorithm would produce lots of missed detections and false alarms. This is the reason behind the development of inherently online anomaly detection methods, as we explain in the next subsection.

7.1.4 Kernel Methods and the Kernel Function

Algorithms based on the so-called kernel trick involve using a *kernel* function that maps the input data onto a *feature space* of much higher dimension [18]. This counterintuitive operation is performed owing to the expectation that points depicting similar behavior should form more pronounced clusters in the richer feature space. A suitable kernel function, when applied to a pair of input vectors, may be interpreted as an inner product in the feature space [18]. This subsequently allows inner products in the feature space (inner products of the *feature vectors*) to be computed without explicit knowledge of the feature vectors themselves, by simply evaluating the kernel function:

$$k\left(x_i, x_j\right) = \left\langle \phi\left(\tilde{x}_i\right), \phi\left(\tilde{x}_j\right) \right\rangle \tag{7.1}$$

where x_i, x_j denotes the input vectors and ϕ represents the mapping onto the feature space. Using kernel functions thus allows simple comparison of higher-order statistics between the input vectors.

Subsequently, a kernel matrix is defined as $K := \left(k\left(x_i, x_j\right)\right)_{i=j=1}^{n}$, where x_i is a set of observation $(x_i \in X, i = 1, \dots, n)$.

The following are some popular kernels [31]:

■ Linear kernel: $k^{\mathrm{lin}}(X, X') := \langle X, X' \rangle$

■ Polynomial kernel of degree p: $k^{\mathrm{pol}}(X, X') := \left(1 + \gamma \langle X, X' \rangle\right)^p$

■ Gaussian/radial kernel: $k^{\mathrm{RBF}}(X, X') := \exp\left(\dfrac{-\|X - X'\|^2}{2\sigma^2}\right)$

■ Negative kernel: $k^{\mathrm{nd}}(X, X') := -\|X - X'\|^\beta$, $\beta \in [0, 2]$

Support vector machines, kernel principal component analysis (KPCA), and kernel regression are some offline algorithms which use kernel methods. Linear and non-linear classifiers employ different kernel methods. For example, RBF (radial basis function) kernel methods are suitable for data samples which are non-linearly dependent, because it maps samples onto higher dimensional space ($d > 2$) in non-linear fashion.

7.1.5 Online Anomaly Detection Methods

Online algorithms have the ability to learn from a newly arriving data instance, without re-training the whole data obtained to-date from initiation. As online algorithms involve real-time operations, the computational and storage complexities of the algorithms (in terms of both time and memory) are required to not grow with time as the size of the whole (to-date) dataset grows, and preferably be small.

7.2 Kernel-Based Online Anomaly Detection (KOAD)

A non-linear, kernel-based least-squares algorithm was initially introduced in 2004 by Engel et al. [19]. Their algorithm took advantage of the kernel trick to perform linear regression in a high-dimensional feature space in order to recursively calculate minimum mean-squared error solution to non-linear least-squares problems.

The kernel recursive least-squares (KRLS) algorithm tries to solve the problem of regularization and computational cost using online constructive sparsification.

This sparsification method only selects data samples which cannot be represented as an appropriate linear combination of selected samples. The KRLS algorithm incrementally builds a *dictionary* (basis) of *approximately* linearly independent samples.

Ahmed et al. [16] proposed an extended version of the kernel-based least-squares algorithm, which they termed the KOAD algorithm. The KOAD algorithm incorporates two thresholds of approximate linear independence, incorporates exponential forgetting to gradually reduce the importance of past observations, and allows the deletion of previous dictionary members to enable the basis set to dynamically remain current. These features were absent in the foundation KRLS algorithm of Engel et al. [19].

The postulate of Ahmed et al. is that if data points $\{\mathbf{x}_t\}_{t=1:T}$ show normal behavior in the input space, then the corresponding feature vectors $\{\varphi(x_t)\}_{t=1:T}$ will also be expected to cluster. Consequently, it should be possible to explain the region of normality in the feature space using a relatively small *dictionary* of *approximately* linearly independent elements $\left\{\phi\left(\tilde{x}_j\right)\right\}_{j=1:m}$. Feature vector $\varphi(x_t)$ is said to be *approximately* linearly dependent on $\left\{\phi\left(\tilde{x}_j\right)\right\}_{j=1:m}$, with approximation threshold ν, if the projection error δ_t satisfies [19]:

$$\delta_t = \min_a \left\| \sum_{j=1}^{m} a_j \times \Phi\left(\tilde{x}_j\right) - \Phi\left(x_t\right) \right\|^2 < \nu, \tag{7.2}$$

where $a = \{a_j\}_{j=1:m}$ is the optimal coefficient vector. Here, $\left\{\phi\left(\tilde{x}_j\right)\right\}_{j=1:m}$ represents those $\{x_t\}_{t=1:T}$ that are entered into the dictionary. The size of the dictionary m is expected to be much less than total time steps elapsed T, thereby leading to computational and storage savings.

It is observed that (7.2) involves an L2 norm, which may be simplified exclusively in terms of the inner products of $\phi\left(\tilde{x}_j\right)$ and $\phi\left(x_t\right)$, and thus evaluated using the kernel function without explicit knowledge of the feature vectors themselves:

$$\delta_t = \min_a = \left\| a_t^T \tilde{K}_{t-1} a_t - 2a_t^T \tilde{k}_{t-1}\left(x_t\right) + k\left(x_t, x_t\right) \right\|^2 < \nu \tag{7.3}$$

where $\left[\tilde{K}_{t-1}\right]_j = k\left(\tilde{x}_t, \tilde{x}_t\right)$ and $\left[\tilde{K}_{t-1}\left(x_t\right)\right]_j = k\left(\tilde{x}_t, x_t\right)$ for $i, j = 1 \ldots m_{t-1}$. Recall that $k(,)$ denotes the kernel function. The optimum sparsification coefficient vector that minimizes δ_t is then:

$$\tilde{a}_t = \tilde{K}_{t-1}^{-1} \tilde{k}_{t-1}\left(x_t\right) \tag{7.4}$$

The expression for error δ_t may then be simplified into:

$$\delta_t = k_{tt} - \tilde{k}_{t-1}\left(x_t\right)^T \tilde{a}_t \leq \nu \tag{7.5}$$

The KOAD algorithm operates at each time step t on a measurement vector $\mathbf{x}_t$. It begins by evaluating the error δ_t in projecting the arriving observation $\mathbf{x}_t$ onto the current dictionary (in the feature domain). This error measure δ_t is then compared with two thresholds ν_1 and ν_2, where $\nu_1 < \nu_2$. If $\delta_t < \nu_1$, KOAD infers that x_t is sufficiently linearly dependent on the dictionary and thus represents normal behavior. If $\delta_t > \nu_2$, KOAD concludes that x_t is far away from the realm of normality and consequently raises a "Red1" alarm to immediately signal an anomaly.

In the interim case where $\nu_1 < \delta_t < \nu_2$, KOAD infers that x_t is sufficiently linearly independent from the dictionary to be considered an unusual event. It may indeed be an anomaly, or it may represent an expansion or migration of the space of normality itself. In this case, KOAD does the following: It immediately raises an "Orange" alarm, then keeps track of the contribution of the relevant input vector x_t in explaining subsequent arrivals for a further ℓ time steps, and at time $t + \ell$ ultimately resolves x_t into either "Green" alarm meaning normal observation or a "Red2" alarm meaning anomalous observation.

The application realm of the KOAD algorithm has subsequently expanded into other domains such as medical monitoring [20] and surveillance systems [21,22] using image processing techniques [23–25]. It has been cited by more than 100 online anomaly detection studies in the field of image processing, e-health, network traffic analysis, and so on. It has also been selected for a reading list of classic papers for a seminar course at the prestigious Swiss Federal Institute of Technology (ETH) Zurich, Switzerland [17]. This impressive application sphere of the algorithm has inspired us to prepare a tutorial guideline that presents a detailed, step-by-step derivation and development of the core algorithm, along with all the mathematical proofs involved. The expectation is for the detailed derivation and tutorial to aid researchers aiming to theoretically extend the core algorithm itself.

Algorithm 7.1 Kernel-Based Online Anomaly Detection [21]

Algorithm 7.1 presents pseudocode for the KOAD algorithm.

```
1 Set thresholds: ν₁, ν₂;
2 for t = 1, 2, …do
Data: (xₜ, yₜ)
/* Evaluate current measurement */
3      Compute projection error δₜ for xₜ using dictionary Dₜ;
4      if δₜ > ν₂ then
5           Raise Red1_Alarm(xₜ);
6      endif
7      if δₜ > ν₁ then
8           Raise Orange_Alarm(xₜ);
9           Store xₜ in Θ;
10     endif
```

```
/* Process previous orange alarm */
11   if Orange_Alarm(x_{t-l}) then
12          Re-evaluate projection error δ for x_{t-l} using dictionary D_t;
13          if δ > ν_1 then
14                 Evaluate usefulness of x_{t-l} over previous l measurements;
15                 if NOT useful then
16                        Raise Red2_Alarm(x_{t-l});
17                 else
18                        Add x_{t-l} to dictionary D;
19                        Lower Orange_Alarm(x_{t-l});
20                 endif
21          else
22                 Lower Orange_Alarm(x_{t-l});
23          endif
24          Remove Θ{1};
25   endif
/* Remove obsolete elements */
26   Evaluate usefulness of each dictionary element over previous L
     measurements;
27   Remove any useless element from dictionary D;
28 endfor
```

7.2.1 KOAD Algorithm Description

This section provides an overview of KOAD, before the algorithm is presented in detail in Section 7.3. In this section, we introduce the parameters of the algorithm and the ways that they are set. We also provide some numeric examples to visualize a schema of the various attributes, characteristics, and parameters of the algorithm during compilation.

7.2.2 Set Thresholds v_1, v_2

This algorithm does not yet provide automatic setting of the thresholds, ν_1 and ν_2. Ahmed et al. [21] selected 300 training samples for their experiment and set v_2 to one for the training period. They also investigated different pairs of v_1 and v_2 and showed that the optimal setting varies for different metrics. The authors also concluded that the performance of different settings remains the same for the same metrics. We recommend that researchers run the algorithm over a training dataset in a supervised fashion with pre-known anomalies and then, set the threshold values that result in an acceptable compromise between detection and false alarm rates.

7.2.3 *Choose Parameters l, ε, L, L, d, γ*

- *l*: This is a parameter for resolving orange alarm. When $v_1 < \delta_t < v_2$, the algorithm waits for l time steps and then decides whether to elevate the existing orange alarm to red, or to add the corresponding vector input to the dictionary. Adding the vector to the dictionary indicates a change in the basis for the sphere of normality. l should be selected in a manner which balances the waiting time to detect anomaly and the false alarm rate. If it is selected as too long, it will violate the principle of real-time detection. If it is selected too small, there will not be enough time to make an intelligent decision for a gray case.

- *ε*: This is also a parameter for resolving orange alarm. Parameter $\varepsilon \in (0, 1)$. It determines what fraction of input vectors should lie within the region of usefulness. The algorithm should be run with different value of ε, and the effect of ε on performance should be investigated. It should be selected based on the user's sensitivity tolerance.

- *L*: This is a parameter for dropping obsolete elements. It determines the time when obsolete (useless) elements are removed from the dictionary. This should be set based on the long-term stationarity of the application data sphere.

- *d*: This is also a parameter for dropping obsolete elements. It determines the amount of *closeness* between a dictionary element and an input vector in order to consider a dictionary element as useful. In other words, it defines the region of *usefulness*. The value of d should be selected based on the kernel type and value, because the kernel implicitly defines a distance measure. The value of d should be below the average kernel value of any genuine dictionary element.

- *γ*: This is the forgetting factor. The algorithm gradually disregards old data in an exponential manner. Parameter γ is a time-based weight which is systematically applied to old observations. A value of $\gamma = 1$ means that recent and previous input vectors have equal importance. The forgetting factor is set γ^n ($0 < \gamma < 1$, $n = 1, 2, 3, \ldots$) for the nth most-recent observation, meaning that recent events are gradually more important than past events.

7.2.4 *Initialization Phase*

- $t = 1$, $\mathcal{D} = \{x_1\}$, $m_1 = 1$, $\tilde{K}_1 = [k_{11}]$, $\tilde{K}_1^{-1} = \begin{bmatrix} \frac{1}{k_{11}} \end{bmatrix}$, $\tilde{\alpha}_1 = \frac{y_1}{k_{11}}$, $P_1 = [1]$, $\Lambda = [1]$

- $\mathcal{D} = \{x_1\}$: The first input vector is added to the dictionary.

- $m_1 = 1$: The number of elements in dictionary (in correspondence with preceding step) is one.

- $\tilde{K}_1 = [k_{11}]$: The kernel matrix is set to the kernel value of the (as of now) sole element of the dictionary with itself. In general, $\tilde{K}_t$ keeps track of kernel values among the members of the dictionary at time t.

■ Example, assume that $t = 5$ and $\mathcal{D} = \{x_1, x_3, x_5\}$, $x_1 = dic_1$, $x_3 = dic_2$, and $x_5 = dic_3$. That is, the first, third, and fifth arriving samples have been entered into the dictionary, with a total of 5 time steps having elapsed since the algorithm began running, and thereby constitute the dictionary composition at time $t = 5$. Then:

■ $$\tilde{K}_5 = \begin{bmatrix} k(dic_1, dic_1) = 1 & k(dic_1, dic_2) & k(dic_1, dic_3) \\ k(dic_2, dic_1) & k(dic_2, dic_2) = 1 & k(dic_2, dic_3) \\ k(dic_3, dic_1) & k(dic_3, dic_2) & k(dic_3, dic_3) = 1 \end{bmatrix}$$

■ $\tilde{K}_1^{-1} = \left[\dfrac{1}{k_{11}}\right]$: The inverse of kernel matrix.

■ $\tilde{\alpha}_1 = \dfrac{y_1}{k_{11}}$: The coefficient least-squares vector α at $t = 1$.

■ $P_1 = [1]$: **P** is the covariance matrix and equal to $\left[A^T A\right]^{-1}$.

■ $A_t = []_{t \times m}$: Is a matrix of least-squares coefficients $a = (a_1, a_2, \ldots, a_m)$.
  – Example 1: Assume that $t = 7$ and $\mathcal{D} = \{x_1, x_3, x_5\}$. Then:

$$A = \begin{bmatrix} a_{11} = 1 & a_{13} & a_{15} \\ a_{21} & a_{23} & a_{25} \\ a_{31} & a_{33} = 1 & a_{35} \\ a_{41} & a_{43} & a_{45} \\ a_{51} & a_{53} & a_{55} = 1 \\ a_{61} & a_{63} & a_{65} \\ a_{71} & a_{73} & a_{75} \end{bmatrix}.$$

  – Example 2: x_4 and x_7 can be shown as below:

$$x_4 = a_{41} x_1 + a_{43} x_3$$

$$x_7 = a_{71} x_1 + a_{73} x_3 + a_{75} x_5$$

■ In order to obtain a recursive formula for $\mathbf{P}_t$, the matrix inversion lemma [26] is used.
■ Λ: This is a binary matrix. It concatenates two submatrices of sizes $L \times m_{t-1}$ (#columns is equal to the number of dictionary members in time $t - 1$) and $L \times G$ (#columns is equal to the number of unsolved orange alarms). It keeps track of whether kernel values of x_t with each dictionary member, and kernel

values of x_t with each of unsolved orange alarm, exceed the value of d for the previous L time steps or not.

7.2.5 Calculate Projection Error δ_t

■ For each arriving input vector "x" at time "t," the projection error "δ_t" should be evaluated:
$$\delta_t = k_{tt} - \tilde{k}_{t-1}(x_t)^T \cdot a_t \text{ where } k_{tt} = k(x_t, x_t).$$

7.2.6 Compute $\tilde{k}_{t-1}(x_t)$

■ The first step to evaluate "δ_t" is the computation of the kernel values for the current input vector.
■ The vector $\tilde{k}_{t-1}(x_t)$ is the kernel value of the current input vector with each dictionary element.
   − Example: Assume that $t = 5$ and $\mathcal{D} = \{x_1 x_3, x_5\}$. Then $\tilde{k}_4(x_5) =$
$$\begin{bmatrix} k(dic_1, x_5) \\ k(dic_2, x_5) \\ k(dic_3, x_5) \end{bmatrix}.$$

7.2.7 Compute Sparsification Vector a_t

■ $a_t = \tilde{K}_{t-1}^{-1} \cdot \tilde{k}_{t-1}(x_t)$

7.2.8 Update Λ

■ If $t > L$: Remove first row of matrix Λ and append Λ with 1 or 0 (1: When kernel values of x_t with each of dictionary members exceed the value of d; 0: When kernel values of x_t with each of dictionary members does not exceed the value of d).
■ If $t < L$: Append Λ with one or zero.

7.2.9 Raise Red1 Alarm

■ If $\delta_t > v_2$
■ Only "A" changes between time steps, and $\tilde{K}_t$ remains unchanged.
■ The Red1 alarm will be raised. It means that the current input vector is significantly different from the pattern of data supported by the dictionary members, and the input vector is consequently marked as an anomaly.

7.2.10 Raise Orange Alarm

■ If $\delta_t > v_1$ & $\delta_t < v_2$:

■ Set $\Theta = [\Theta \cup x_t]$, $\mathcal{D} = [\mathcal{D} \cup x_t]$, where Θ is the set of unsolved Orange alarms $\tilde{A}_t = a_t$; compute $\tilde{K}_1^{-1}$ and $\tilde{K}_t$; append Λ with $(0,\dots,1)^T$.

■ $\tilde{K}_1^{-1} = \begin{bmatrix} \delta_t \tilde{K}_{t-1}^{-1} + \tilde{a}_t \tilde{a}_t^T & -\tilde{a}_t \\ -\tilde{a}_t^T & 1 \end{bmatrix}$

■ $\tilde{K}_t = \begin{bmatrix} \tilde{K}_{t-1} & \tilde{k}_{t-1}(x_t) \\ \tilde{k}_{t-1}(x_t)^T & k_{tt} \end{bmatrix}$

– Example 1: Assume $t = 5$, $\mathcal{D} = \{x_1, x_5\}$, and $\Theta = \{x_5\}$. Then, $\Lambda = \begin{bmatrix} 1 & 0 \\ 1 & 0 \\ 1 & 0 \\ 1 & 0 \\ 0 & 1 \end{bmatrix}$. The second column is the result of $(0\ 1)^T$ at time $t = 5$ when

the orange alarm is raised.

– Example 2: Assume $t = 10$, $\mathcal{D} = \{x_1, x_5, x_{10}\}$, $l = 20$ and $\Theta = \{x_5, x_{10}\}$. Then

$$\Lambda = \begin{bmatrix} 1 & 0 & 0 \\ 1 & 0 & 0 \\ 1 & 0 & 0 \\ 1 & 0 & 0 \\ 0 & 1 & 0 \\ 1 & 0 & 0 \\ 1 & 0 & 0 \\ 1 & 0 & 0 \\ 1 & 0 & 0 \\ 0 & 0 & 1 \end{bmatrix}$$. Both x_5, x_{10} are unsolved orange alarms.

– Example: Assume that $t = 5$, $m = 2$, $\mathcal{D} = \{x_1, x_3\}$, $x_1 = dic_1$, $x_3 = dic_2$, and $\delta_5 < v_2$. Then, x_5 causes the orange alarm. Observation x_5 is not linearly dependent on the dictionary elements. Therefore, x_5 cannot be expressed in the form of: $a_1 \times dic_1 + a_2 \times dic_2$. Observation x_5 can then be stated as: $0 \times dic_1 + 0 \times dic_2 + 1 \times x_5$.

The corresponding coefficient vector a_5 when x_5 is added to the dictionary is thus:

$$a_5 = \begin{bmatrix} 0 \\ 0 \\ 1 \end{bmatrix}.$$

7.2.11 Compute P_t

■ $P_t = \dfrac{1}{\gamma}\begin{pmatrix} P_{t-1} & 0 \\ 0^T & \gamma \end{pmatrix}$

- Example: Assume $t = 5$, $m = 2$, $\mathcal{D} = \{x_1, x_3\}$, and $\delta_5 < v_2$. Then, P_5 will be equal to:

$$\begin{bmatrix} P_{t-1} & & 0 \\ & & 0 \\ 0 & 0 & 1 \end{bmatrix}.$$

7.2.12 Compute α_t

■ $\alpha_t = \begin{bmatrix} \gamma^{\frac{-1}{2}} \times \tilde{\alpha}_{t-1} - \dfrac{\tilde{\alpha}_t}{\delta_t}\left(y_t - \gamma^{\frac{-1}{2}} \times \tilde{k}_{t-1}(x_t)^T \times \tilde{\alpha}_{t-1}\right) \\ \dfrac{1}{\delta_t}\left(y_t - \gamma^{\frac{-1}{2}} \times \tilde{k}_{t-1}(x_t)^T \times \tilde{\alpha}_{t-1}\right) \end{bmatrix}$

■ As the current input vector is added to the dictionary, the size of dictionary is incremented.

7.2.13 Compute q, P_t, and α_t When $\delta_t > v_2$

■ When $\delta_t > v_2$, matrix A does not change between time steps and $\tilde{K}_t = \tilde{K}_{t-1}$ (i.e., kernel matrix remains unchanged).

■ $q_t = \dfrac{P_{t-1}\alpha_t}{\gamma + a_t^T P_{t-1}\alpha_t}$

■ $P_t = \dfrac{1}{\gamma}\left(P_{t-1} - q_t a_t^T P_{t-1}\right)$

■ $\tilde{\alpha}_t = \tilde{\alpha}_{t-1} + \tilde{K}_{t-1}^{-1} q_t \left(y_t - \tilde{k}_{t-1}(x_t)^T \tilde{\alpha}_{t-1}\right).$

7.2.14 Lower Orange Alarm (x_{t-1}) to Green

■ If there is an unsolved orange alarm at time step $t - 1$, we need to perform a secondary "Usefulness Test" to resolve the orange alarm. A dictionary element "x_{t-1}" is regarded as useful if it was used to explain a *significant number* of input vectors between time steps $t - l$ to t. In other words, if a noticeable number of kernel values between x_{t-l} and $(x_{t-l+1}, x_{t-l+2}, ..., x_t)$ are high, then x_{t-l} should be added to the dictionary, and subsequently, x_{t-l} should not be considered anomalous.

- In the above situation, the orange alarm is reduced to green. It demonstrates the migration or expansion of normal traffic in the feature space.
- As the $(m_{t-1}+1)$th column of matrix Λ keeps track of kernel values of the x_{t-l} (orange alarm) with $(x_{t-l+1}, x_{t-l+2}, ..., x_t)$, we need to investigate the matrix Λ.
- KOAD evaluates the sum of all the kernel values between x_{t-1} and $(x_{t-l+1}, x_{t-l+2}, ..., x_t)$ and compares whether it is less than a specific value or not. If it is less than $(\varepsilon \times l)$, then it will be considered anomalous and the orange alarm will be elevated to a Red2 alarm.
  - Example 1: Assume $t = 11$, $\mathcal{D} = \{x_1, x_4, x_{10}\}$, $l = 7$ and $\Theta = \{x_4, x_7\}$,

$$\varepsilon = 0.2, \text{ and } \Lambda = \begin{bmatrix} 1 & 0 & 0 \\ 1 & 0 & 0 \\ 1 & 0 & 0 \\ 0 & 1 & 0 \\ 1 & 0 & 0 \\ 1 & 0 & 0 \\ 1 & 0 & 1 \\ 1 & 0 & 0 \\ 1 & 0 & 1 \\ 1 & 0 & 1 \\ 0 & 0 & 1 \end{bmatrix}. \text{ For resolving orange alarm, we need}$$

to compute sum (Λ (5:11,2)), which evaluates to zero as $0 < (0.2 \times 7)$. Therefore, the orange alarm should be elevated to Red2 alarm.

7.2.15 Remove Obsolete Elements

- When the kernel value of x_{t-L} and all incoming input vectors up to x_t become zero, it causes the relevant column of Λ to contain all zeros. As a result, the x_{t-L}th member of the dictionary will be marked obsolete and should be removed.

7.2.16 Drop pth Element from Dictionary

- This needs to be done either when a previous orange alarm is upgraded to Red2 alarm or when a dictionary element becomes obsolete.

▪ First, move the pth row and columns of $\tilde{K}_t$ and $\tilde{K}_t^{-1}$ to the end. Then, the kernel values of every other elements with pth element will be associated with last row and column of $\tilde{K}_t$ and $\tilde{K}_t^{-1}$.

7.2.17 Calculate δ_p and $\tilde{a}_p$

▪ $\delta_p = \dfrac{1}{\left[\tilde{K}_t^{-1}\right]_{m_t,m_t}}$

▪ $\tilde{a}_p = -\delta_p \times \left[\tilde{K}_t^{-1}\right]_{1:m_t-1,m_t}$.

7.2.18 Calculate $\tilde{K}_t^{-1}$ and $\tilde{a}_t$

▪ $\tilde{K}_t^{-1} = \left[\tilde{K}_t^{-1}\right]_{1:m_t-1,m_t-1} - \dfrac{\tilde{a}_p \tilde{a}_p^T}{\delta_p}$

▪ $\tilde{\alpha}_t = \tilde{\alpha}_t - \dfrac{1}{\delta_p}\begin{pmatrix} \tilde{a}_p \tilde{a}_p^T & -\tilde{a}_p \\ -\tilde{a}_p^T & 1 \end{pmatrix}\tilde{K}_t\tilde{\alpha}_t .$

7.2.19 Set a_t and $\tilde{K}_t$

▪ $a_t = a_t(1:m-1)$

▪ $\tilde{K}_t = \left[\tilde{K}_t\right]_{1:m_t-1,m_t-1}$.

7.2.20 Update $\mathcal{D}$, Λ, m, and P

▪ Remove pth element from $\mathcal{D}$.
▪ Remove pth column from Λ.
▪ Set $m_t = m_{t-1} - 1$.
▪ Set $P = C \times I_{m_t}$
  − The recalculation of the covariance matrix P requires full access to historical data. In order to simplify the calculation, the matrix P is rest to a large constant C times the m_t-sized identity matrix.
  − Example, if $C = 10,000$ and $m_t = 3$, then

$$P = 10,000 \times \begin{bmatrix} 1 & 0 & 0 \\ 0 & 1 & 0 \\ 0 & 0 & 1 \end{bmatrix}.$$

7.2.21 Compute $\tilde{K}_{t-1}$

■ This matrix contains the kernel values of all x_{t-1} dictionary members between themselves.
  – Example: For $m = 6$,

$$\tilde{K}_{t-1} = \begin{bmatrix} k(dic_1 \text{ and } x_{t-1}) \\ k(dic_2 \text{ and } x_{t-1}) \\ k(dic_3 \text{ and } x_{t-1}) \\ k(dic_4 \text{ and } x_{t-1}) \\ k(dic_5 \text{ and } x_{t-1}) \\ k(dic_6 \text{ and } x_{t-1}) \end{bmatrix}.$$

7.3 Mathematical Proofs of KOAD Algorithm

In this section, we go traverse through the derivations step-by-step. We believe that the detailed steps of the mathematical derivations should be recorded, as it will aid future researchers in extending and expanding our work.

7.3.1 Formalizing δ

■
$$\left\| \sum_{j=1}^{m(t-1)} a_j \cdot \phi(\tilde{x}_j) - \phi(x_t) \right\|^2$$

$$= \left\langle \left\{ \sum_{j=1}^{m(t-1)} a_j \cdot \phi(\tilde{x}_j) - \phi(x_t) \right\}, \left\{ \sum_{j=1}^{m(t-1)} a_j \cdot \phi(\tilde{x}_j) - \phi(x_t) \right\} \right\rangle$$

$$= \sum_{i=1}^{m(t-1)} \sum_{j=1}^{m(t-1)} a_i a_j \left\langle \phi(\tilde{x}_i), \phi(\tilde{x}_j) \right\rangle - \left\langle \phi(x_t), \sum_{j=1}^{m(t-1)} a_j \cdot \phi(\tilde{x}_j) \right\rangle$$

$$- \left\langle \phi(x_t), \sum_{j=1}^{m(t-1)} a_j \cdot \phi(\tilde{x}_j) \right\rangle + \left\langle \phi(x_t), \phi(x_t) \right\rangle$$

$$= \sum_{i=1}^{m(t-1)} \sum_{j=1}^{m(t-1)} a_i a_j \left\langle \phi(\tilde{x}_i), \phi(\tilde{x}_j) \right\rangle$$

$$- 2 \sum_{j=1}^{m(t-1)} a_j \left\langle \phi(\tilde{x}_j), \phi(x_t) \right\rangle + \left\langle \phi(x_t), \phi(x_t) \right\rangle$$

$$\boxed{\delta_t = \min_a \left\{ a^T \tilde{K} a - 2a^T \tilde{k}(x_t) + k_{tt} \right\}}$$

where $\tilde{K} = \left[\tilde{K}_{t-1} \right]_{ij} = \left[k(\tilde{x}_i, \tilde{x}_j) \right]$, $i, j = 1, \ldots, m(t-1)$

- $\tilde{k}(x_t) = \left[\tilde{k}(x_t)\right]_i = \left[k(\tilde{x}_i, x_t)\right], i = 1, \ldots, m(t-1)$
- $k_{tt} = k(x_t, x_t)$
- Example: Let $m(t-1) = 3$. Then:

$$a^T \tilde{K} a = \begin{bmatrix} a_1 & a_2 & a_3 \end{bmatrix} \begin{bmatrix} k_{11} & k_{12} & k_{13} \\ k_{21} & k_{22} & k_{23} \\ k_{31} & k_{32} & k_{33} \end{bmatrix} \begin{bmatrix} a_1 \\ a_2 \\ a_3 \end{bmatrix}$$

$$= \begin{bmatrix} a_1 k_{11} + a_2 k_{21} + a_3 k_{31} & a_1 k_{12} + a_2 k_{22} + a_3 k_{32} & a_1 k_{13} + a_2 k_{23} + a_3 k_{33} \end{bmatrix} \cdot \begin{bmatrix} a_1 \\ a_2 \\ a_3 \end{bmatrix}$$

$$= a_1 a_1 k_{11} + a_1 a_2 k_{21} + a_1 a_3 k_{31} + a_2 a_1 k_{12} + a_2 a_2 k_{22} + a_2 a_3 k_{32} + a_3 a_1 k_{13} + a_3 a_2 k_{23}$$
$$+ a_3 a_3 k_{33} \tag{7.6}$$

On the other hand,

$$\sum_{i=1}^{m(t-1)} \sum_{j=1}^{m(t-1)} a_i a_j \cdot \left\langle \phi(\tilde{x}_i), \phi(\tilde{x}_j) \right\rangle = \sum_{i=1}^{3} \sum_{j=1}^{3} a_i a_j \, k(\tilde{x}_i, \tilde{x}_j)$$

$$= \sum_{i=1}^{3} a_i a_1 \, k(\tilde{x}_i, \tilde{x}_1) + a_i a_2 \, k(\tilde{x}_i, \tilde{x}_2) + a_i a_3 \, k(\tilde{x}_i, \tilde{x}_3)$$

$$= a_1 a_1 k_{11} + a_1 a_2 k_{12} + a_1 a_3 k_{13} + a_2 a_1 k_{21} + a_2 a_2 k_{22} + a_2 a_3 k_{23} + a_3 a_1 k_{31} + a_3 a_2 k_{32} + a_3 a_3 k_{33}$$
$$\tag{7.7}$$

As $(7.6) = (7.7)$:

$$a^T \tilde{K} a = \sum_{i=1}^{m(t-1)} \sum_{j=1}^{m(t-1)} a_i a_j \left\langle \phi(\tilde{x}_i), \phi(\tilde{x}_j) \right\rangle$$

- Example: Let $m(t-1) = 3$. Then:

$$a^T \tilde{K}(x_t) = \begin{bmatrix} a_1 & a_2 & a_3 \end{bmatrix} \begin{bmatrix} k(\tilde{x}_1, x_t) \\ k(\tilde{x}_2, x_t) \\ k(\tilde{x}_3, x_t) \end{bmatrix} \tag{7.8}$$

$$= a_1 k(\tilde{x}_1, x_t) + a_2 k(\tilde{x}_2, x_t) + a_3 k(\tilde{x}_3, x_t)$$

$$\sum_{j=1}^{m(t-1)} a_j \langle \phi(\tilde{x}_j), \phi(x_t) \rangle$$

$$= \sum_{j=1}^{3} a_j \cdot k(\tilde{x}_j, x_t) \qquad (7.9)$$

$$= a_1 k(\tilde{x}_1, x_t) + a_2 k(\tilde{x}_2, x_t) + a_3 k(\tilde{x}_3, x_t)$$

As (7.8) = (7.9):

$$a^T \tilde{K}(x_t) = \sum_{j=1}^{m(t-1)} a_j \langle \phi(\tilde{x}_j), \phi(x_t) \rangle.$$

7.3.2 Formalizing a

■ $\delta_t = \min_a \left\{ a^T \tilde{K} a - 2a^T \tilde{k}(x_t) + k_{tt} \right\}$

■ So, $\dfrac{\partial}{\partial a}(\delta_t) = 0^*$

  $- \dfrac{\partial}{\partial a}(a^T \tilde{K} a) - 2\dfrac{\partial}{\partial a}\left\{ a^T \tilde{k}(x_t) \right\} + \dfrac{\partial}{\partial a}\left\{ k_{tt} \right\} = 0$

  $- \left[\dfrac{\partial}{\partial a}\left\{ a^T (\tilde{K}a) \right\} + \dfrac{\partial}{\partial a}\left\{ (a^T \tilde{K})a \right\} \right] - 2\tilde{k}(x_t) + 0 = 0$

  $- \left[(\tilde{K}a) + (a^T \tilde{K})^T \right] - 2\tilde{k}(x_t) = 0$

  $- \left[(\tilde{K}a) + (\tilde{K}^T a) \right] = 2\tilde{k}(x_t)$

■ As $\tilde{K}^T = \tilde{K}$ then $\left[(\tilde{K} + \tilde{K})a \right] = 2\tilde{k}(x_t)$

  $- 2\tilde{K}a = 2\tilde{k}(x_t)$

  $- \tilde{K}a = \tilde{k}(x_t)$

■ Multiplying both sides with $\tilde{K}^{-1}$

  $- \tilde{K}^{-1}\tilde{K}a = \tilde{K}^{-1}\tilde{k}(x_t)$

  $- Ia = \tilde{K}^{-1}\tilde{k}(x_t)$

$$\boxed{\tilde{a} = \tilde{K}_{t-1}^{-1}\tilde{k}_{t-1}(x_t)}$$

■ Therefore, $\tilde{a} = \tilde{K}_{t-1}^{-1}\tilde{k}_{t-1}(x_t)$ is the value of that minimizes δ_t.

■ To compute the value of δ_t when $\tilde{a} = \tilde{K}_{t-1}^{-1}\tilde{k}_{t-1}(x_t)$:

  $- \delta_t = a^T \tilde{K} a - 2a^T \tilde{k}(x_t) + k_{tt}$ when

$$a = \tilde{a} = \tilde{K}_{t-1}^{-1}\tilde{k}_{t-1}(x_t)$$

$$= \left[\tilde{K}^{-1}\tilde{k}(x_t)\right]^T \tilde{K}\left[\tilde{K}^{-1}\tilde{k}(x_t)\right] - 2\left[\tilde{K}^{-1}\tilde{k}(x_t)\right]^T \tilde{k}(x_t) + k_{tt}$$

$$= \left[\tilde{k}(x_t)^T \cdot \tilde{K}^{-T}\right] \cdot \left[\tilde{K}\tilde{K}^{-1}\right] \cdot \tilde{k}(x_t) - 2\tilde{k}(x_t)^T \cdot \tilde{K}^{-T}\tilde{k}(x_t) + k_{tt}$$

$$= \tilde{k}(x_t)^T \cdot \tilde{K}^{-T} \cdot \tilde{k}(x_t) - 2\tilde{k}(x_t)^T \cdot \tilde{K}^{-T}\tilde{k}(x_t) + k_{tt}$$

- As $\tilde{K}^{-T} = \tilde{K}^{-1}$ then

$$\delta_t = \tilde{k}(x_t)^T \cdot \tilde{K}^{-T} \cdot \tilde{k}(x_t) - 2\tilde{k}(x_t)^T \cdot \left[\tilde{K}^{-T}\tilde{k}(x_t)\right] + k_{tt}$$

$$= \tilde{k}(x_t)^T \cdot [\tilde{a}] - 2\tilde{k}(x_t)^T \cdot [\tilde{a}] + k_{tt}$$

$$= -\tilde{k}(x_t)^T \cdot \tilde{a} + k_{tt}$$

$$= -\tilde{k}(x_t)^T \cdot \tilde{a}_t + k_{tt}$$

$$= k_{tt} - \tilde{k}(x_t)^T \cdot \tilde{a}_t$$

7.3.3 Expanding Current Dictionary by Adding x_t

- If $\delta_t > v_1$, then we must expand the current dictionary by inserting x_t in it.
- $\phi(x_t)$ may now be exactly represented using the expanded dictionary, so δ_t is now zero.
- As a result, for every time step up to t, we have:
- $\phi(x_i) = \sum_{j=1}^{m_i} a_{ij}\phi(\tilde{x}_j) + \delta_t$
- Here, $\mathcal{D} = \{\tilde{x}_1, \tilde{x}_2, \ldots, \tilde{x}_{m(i)}\}$, $\delta_t = \phi_i^{res}$, and $\left\|\phi_i^{res}\right\|^2 = \left\|\delta_t\right\|^2 \le v$.
- Then, $\left\langle \phi(x_i), \phi(x_j) \right\rangle \ge \left\langle \sum_{j=1}^{m_i} a_{ij}\phi(\tilde{x}_j), \phi(x_i) \right\rangle$
- Assuming $\delta_t = 0$
  - $K \approx \left\langle \sum_{j=1}^{m_i} a_{ij} \cdot \phi(\tilde{x}_j), \phi(x_i) \right\rangle$
- $K \approx A\tilde{K}A^T$
  where $K = [K_t]_{ij} = k(x_i, x_j)$ with $i, j = 1, 2, \ldots, t$ is the full kernel matrix and $A = [A_T]_{i,j} = a_{ij}$.
- As the algorithm is sequential, $[A_t]_{i,j} = 0$ for $j > m$.
- In practice, we will freely make the substitution $K = A\tilde{K}A^T$ with the understanding that the resulting expressions are approximate whenever v is small and greater than zero.

7.3.4 Properties of the Sparification Method

■ $\Phi = \tilde{\Phi}A^T + \tilde{\Phi}^{res}$
■ Defining the matrices $\Phi = \left[\phi(x_1),\ldots,\phi(x_t)\right]$, $\tilde{\Phi} = \left[\phi(\tilde{x}_1),\ldots,\phi(\tilde{x}_m)\right]$, and $\Phi^{res} = \left[\phi_1^{res},\ldots,\phi_t^{res}\right]$, we can write (7.6) for all samples, as follows:

$$\Phi = \tilde{\Phi}A^T + \Phi^{res} \tag{7.10}$$

■ Pre-multiplying (7.10) by its transpose:

$$\Phi^T\Phi = \left(\tilde{\Phi}A^T + \Phi^{res}\right)^T \left(\tilde{\Phi}A^T + \Phi^{res}\right)$$

■ $K = \tilde{\Phi}^T A \cdot \tilde{\Phi}A^T + \tilde{\Phi}^T A\Phi^{res} + \Phi^{resT}\tilde{\Phi}A^T + \Phi^{resT}\Phi^{res}$
■ Φ^{res} and $\tilde{\Phi}$ are perpendicular vectors, so $\Phi^{res} \cdot \tilde{\Phi} = 0$
  – $K = \tilde{\Phi}^T A\tilde{\Phi}A^T + \Phi^{resT}\Phi^{res}$
■ As $R = \Phi^{resT}\Phi^{res}$ (R: residual)
  – $K = \tilde{\Phi}^T A\tilde{\Phi}A^T + R$
■ Assuming $R \approx 0$

$$K \approx \tilde{\Phi}^T A\tilde{\Phi}A^T$$

$$K = A\tilde{K}A^T \tag{7.11}$$

7.3.5 The Kernel Recursive Least-Squares Algorithm

■ In the simplest form of the recursive least-squares (RLS) algorithm, the sum of the following squared errors is minimized at each time step t:

$$\mathcal{L}(w) = \sum_{i=1}^t \left(\hat{f}(x_i) - y_i\right)^2 = \left\|\Phi_t^T w - y_t\right\|^2.$$

■ Φ_t^T is a matrix which includes all vectors of data in the dictionary.
■ w is a weight vector.
■ $y_t = (y_1, y_2, \ldots, y_t)^T$.
■ In order to minimize $\mathcal{L}(w)$ with respect to w:
  – $w_t = \text{argmin}_w \left\|\Phi_t^T w - y_t\right\|^2 = \left(\Phi_t^T\right)^+ y_t$
  where $\left(\Phi_t^T\right)^+$ is the pseudo-inverse[†] of Φ_t^T.
■ We may express the optimal weight vector as

$$w_t = \sum_{i=1}^t \alpha_i \phi(x_i) = \Phi_t \alpha \tag{7.12}$$

where $\alpha = (\alpha_1, \alpha_2, \ldots, \alpha_t)^T$.

■ Subsequently, $\mathcal{L}(w)$ may be written as:

$$\mathcal{L}(\alpha) = \left\| \mathbf{K}_t \alpha - y_t \right\|^2 .$$

■ In order to minimize $\mathcal{L}(\alpha) \Rightarrow \left\| \mathbf{K}_t \alpha - y_t \right\|^2 = 0$, but the size of vector α becomes huge when t is large. Therefore, in calculating the weight vector w, we only use the samples in the dictionary.

■ By substituting (7.11) into (7.12)

  – $w_t = \mathbf{\Phi}\alpha_t \approx \tilde{\mathbf{\Phi}}_t A_t^T \alpha_t = \tilde{\mathbf{\Phi}}_t \tilde{\alpha}_t$, where $\tilde{\alpha}_t \overset{\text{def}}{=} A_t^T \alpha_t$ is a vector of m reduced coefficients.

■ The loss becomes $\mathcal{L}(\tilde{\alpha}) = \left\| \mathbf{\Phi}_t^T \tilde{\mathbf{\Phi}} \tilde{\alpha} - y_t \right\|^2 = \left\| A_t \tilde{K}_t \tilde{\alpha} - y_t \right\|^2$.

■ For minimization, $\left\| A_t \tilde{K}_t \tilde{\alpha} - y_t \right\|^2 = 0$.

$$\boxed{\tilde{\alpha}_t = \tilde{K}_t^{-1} A_t^+ - y_t}$$

7.3.6 Finding Recursive Formula for P

■ $P_t = \left(A_t^T A_t \right)^{-1}$

■ Using matrix inversion lemma[‡] to find recursive formula for P:

$$P_t = P_{t-1} - \frac{P_{t-1} a_t a_t^T P_{t-1}}{1 + a_t^T P_{t-1} a_t} \tag{7.13}$$

where $a_t^T P_{t-1} \cdot a_t = b_t^T \cdot P_{t-1}(1:m-1, 1:m-1) \cdot b_t$

■ Let $q_t = \dfrac{P_{t-1} a_t}{1 + a_t^T P_{t-1} a_t}$

■ Replacing the corresponding part of (7.13) with q_t:

$$P_t = P_{t-1} - q_t \cdot a_t^T P_{t-1}$$

■ Multiplying both sides of equation (7.13) with a_t:

$$P_t \cdot a_t = P_{t-1} a_t - \frac{P_{t-1} a_t a_t^T P_{t-1} \cdot a_t}{1 + a_t^T P_{t-1} a_t}$$

$$= \frac{P_{t-1} a_t \left(1 + a_t^T P_{t-1} a_t \right) - P_{t-1} a_t a_t^T P_{t-1} \cdot a_t}{1 + a_t^T P_{t-1} a_t}$$

$$= \frac{P_{t-1} a_t + P_{t-1} a_t a_t^T P_{t-1} \cdot a_t - P_{t-1} a_t a_t^T P_{t-1} \cdot a_t}{1 + a_t^T P_{t-1} a_t}$$

$$= \frac{P_{t-1} a_t}{1 + a_t^T P_{t-1} a_t}$$

$$= q_t$$

7.3.7 Finding Recursive Formula for α: Proof 1

■ $P_{-t} = \left[A_{-t}^T A_{-t} \right]^{-1}$

$\qquad = \tilde{K}_t^{-1} \cdot [P_t] \cdot \left[A_t^T Y_t \right]$

■ $\tilde{\alpha}_t = \tilde{K}_t^{-1} A_t^{-1} Y_t,$ then

$$\tilde{\alpha}_t = \tilde{K}_t^{-1} \left[P_{t-1} - \frac{P_{t-1} a_t a_t^T P_{t-1}}{1 + a_t^T P_{t-1} a_t} \right] \left[A_{t-1}^T Y_{t-1} + a_t y_t \right]$$

$$= \tilde{K}_t^{-1} \left[P_{t-1} - q_t \cdot a_t^T P_{t-1} \right] \left[A_{t-1}^T Y_{t-1} + a_t y_t \right]$$

$$= \tilde{K}_t^{-1} \left[P_{t-1} A_{t-1}^T Y_{t-1} + P_{t-1} a_t y_t - q_t \cdot a_t^T P_{t-1} A_{t-1}^T Y_{t-1} - q_t \cdot a_t^T P_{t-1} a_t y_t \right]$$

$$= \tilde{K}_t^{-1} \left[P_{t-1} a_t y_t - q_t \cdot a_t^T P_{t-1} A_{t-1}^T Y_{t-1} \right] + \tilde{K}_t^{-1} \left[P_{t-1} A_{t-1}^T Y_{t-1} - q_t \cdot a_t^T P_{t-1} a_t y_t \right]$$

$$= \tilde{K}_t^{-1} \left[P_{t-1} A_{t-1}^T Y_{t-1} \right] + \tilde{K}_t^{-1} \left[P_{t-1} a_t y_t - q_t \cdot a_t^T \left(P_{t-1} A_{t-1}^T Y_{t-1} \right) - q_t \cdot a_t^T P_{t-1} a_t y_t \right]$$

$$\tag{7.14}$$

■ When $\delta_t < v$. In this case, $\mathcal{D}_t = \mathcal{D}_{t-1}$; hence, $m_t = m_{t-1}$ and $\tilde{K} = \tilde{K}_{t-1}$.

$$= \tilde{K}_{t-1}^{-1} \left[P_{t-1} - A_{t-1}^T Y_{t-1} \right]$$

$$\quad + \tilde{K}_t^{-1} \left[P_{t-1} a_t y_t - q_t \cdot a_t^T \left(\tilde{K}_{t-1} \right) \left(\tilde{K}_{t-1}^{-1} A_{t-1}^T Y_{t-1} \right) - q_t \cdot a_t^T P_{t-1} a_t y_t \right]$$

$$= \left[\tilde{K}_t^{-1} P_{t-1} A_{t-1}^T Y_{t-1} \right] + \tilde{K}_t^{-1} \left[P_{t-1} a_t y_t - q_t \cdot a_t^T \left(\tilde{K}_t \right) \left(\tilde{\alpha}_{t-1} \right) - q_t \cdot a_t^T P_{t-1} a_t y_t \right]$$

$$= \tilde{\alpha}_{t-1} + \tilde{K}_t^{-1} \left[\left(P_{t-1} - q_t \cdot a_t^T P_{t-1} \right) a_t y_t - q_t \cdot a_t^T \tilde{K}_t \tilde{\alpha}_{t-1} \right]$$

$$= \tilde{\alpha}_{t-1} + \tilde{K}_t^{-1} \left[\left(P_t a_t \right) y_t - q_t \left(\tilde{K}_t^T a_t \right)^T \tilde{\alpha}_{t-1} \right]$$

■ As $q_t = P_t a_t$, $\tilde{K}_t^T = \tilde{K}_t$, and $\tilde{K}_t = \tilde{K}_{t-1}$

$$= \tilde{\alpha}_{t-1} + \tilde{K}_t^{-1} \left[q_t y_t - q_t \left(\tilde{K}_{t-1} a_t \right)^T \tilde{\alpha}_{t-1} \right]$$

■ As $a_t = \tilde{K}_{t-1}^{-1} \tilde{k}_{t-1} \left(x_t \right)$
■ $\tilde{K}_{t-1} \cdot a_t = \tilde{k}_{t-1} \left(x_t \right)$

$$\boxed{\tilde{\alpha}_t = \tilde{\alpha}_{t-1} + \tilde{K}_t^{-1} \cdot q_t \left[y_t - \tilde{k}_{t-1} \left(x_t \right)^T \cdot \tilde{\alpha}_{t-1} \right]}$$

7.3.8 Finding Recursive Formula for α: Proof 2

- $\mathcal{L}(w) = \|K_t\alpha - Y_t\|^2$
- Now $K_t = A_t\tilde{K}_tA_t^T$, where A_t is $t \times m$ and $\tilde{K}_t$ is $m \times m$ matrix. As a result, K_t is $t \times t$ matrix.
- If $\tilde{K}_t$ becomes $(m-1) \times (m-1)$, we will need the first $(m-1)$ columns of A_t.
  Then, let $A_t = \begin{bmatrix} R_t & U_t \end{bmatrix}$, where R_t is $t \times (m-1)$ and U_t is $t \times 1$. Thus, U_t contains non-zero coefficients (entities in a) of the element being dropped in past time steps.
- Then, $K_t = R_t\tilde{K}_tR_t^T$.
- Note: where elements are dropped, past coefficients of a (in R_t) become non-optimal. Subsequently:
  - $\mathcal{L}(\tilde{\alpha}_t) = \|K_t\alpha - Y_t\|^2$
  - $\mathcal{L}(\tilde{\alpha}_t) = \left\|\left(R_t\tilde{K}_tR_t^T\right)\alpha - Y_t\right\|^2$
  - $\mathcal{L}(\tilde{\alpha}_t) = \left\|R_t\tilde{K}_t\left(R_t^T\alpha\right) - Y_t\right\|^2$
  - As $R_t^T\alpha = \tilde{\alpha}$
    - $\mathcal{L}(\tilde{\alpha}_t) = \left\|R_t\tilde{K}_t\left(\tilde{\alpha}\right) - Y_t\right\|^2$
  - Then, $\dfrac{\partial}{\partial\tilde{\alpha}}\left\{\mathcal{L}\left(\tilde{\alpha}_t\right)\right\} = 0$
  - $\left[R_t\tilde{K}_t\left(\tilde{\alpha}\right) - Y_t\right] = 0$
    - $R_t\tilde{K}_t\tilde{\alpha} = Y_t$
  - Note that here R_t is $t \times (m-1)$, $\tilde{K}_t$ is $(m-1) \times (m-1)$, and $\tilde{\alpha}$ is $(m-1) \times 1$.
  - Then, $\begin{bmatrix} R_t & U_t \end{bmatrix} \cdot \begin{bmatrix} \tilde{K}_t & 0 \\ 0 & 1 \end{bmatrix}\begin{bmatrix} \tilde{\alpha} \\ 0 \end{bmatrix} = \begin{bmatrix} R_t\tilde{K}_t & U_t \end{bmatrix}\begin{bmatrix} \tilde{\alpha} \\ 0 \end{bmatrix} = R_t\tilde{K}_t\tilde{\alpha} = Y_t$
  - For example, we can write $Y_t = A_t\begin{bmatrix} \tilde{K}_t & 0 \\ 0 & 1 \end{bmatrix}\begin{bmatrix} \tilde{\alpha} \\ 0 \end{bmatrix}$ whenever K_t is $(m-1) \times (m-1)$ and $Y_t = A_t\tilde{K}_t\tilde{\alpha}$, where K_t is $m \times m$ and $\tilde{K}_t$ is $(m-1) \times (m-1)$
    - $Y_t = A_t\begin{bmatrix} \tilde{K}_t & 0 \\ 0 & 1 \end{bmatrix}\begin{bmatrix} \tilde{\alpha} \\ 0 \end{bmatrix}$
    - $A_t^+Y_t = \begin{bmatrix} \tilde{K}_t & 0 \\ 0 & 1 \end{bmatrix}\begin{bmatrix} \tilde{\alpha} \\ 0 \end{bmatrix}$
  - where A^+ is the pseudo-inverse matrix of A_t, as A_t is $t \times m$ and not a square matrix.

$$\bullet \begin{bmatrix} \tilde{K}_t & 0 \\ 0 & 1 \end{bmatrix}^{-1} A_t^+ Y_t = \begin{bmatrix} \tilde{\alpha} \\ 0 \end{bmatrix}$$

$$\bullet \begin{bmatrix} \tilde{K}_t^{-1} & 0 \\ 0 & 1 \end{bmatrix} \cdot \left(P_t A_t^T \right) Y_t = \begin{bmatrix} \tilde{\alpha} \\ 0 \end{bmatrix}$$

- Where $P_t = \left[A_t^T \cdot A_t \right]^{-1}$, so $A_t^+ = P^T A_t^T$

$$- \begin{bmatrix} \tilde{\alpha} \\ 0 \end{bmatrix} = \begin{bmatrix} \tilde{K}_t^{-1} & 0 \\ 0 & 1 \end{bmatrix} P_t A_t^T Y_t$$

$$- \begin{bmatrix} \tilde{\alpha} \\ 0 \end{bmatrix} = \begin{bmatrix} \tilde{K}_t^{-1} & 0 \\ 0 & 1 \end{bmatrix} \left[P_t \right] \left[A_t^T Y_t \right]$$

$$= \begin{bmatrix} \tilde{K}_t^{-1} & 0 \\ 0 & 1 \end{bmatrix} \left[P_{t-1} - \frac{P_{t-1} a_t a_t^T P_{t-1}}{1 + a_t^T P_{t-1} a_t} \right] \cdot \left[A_{t-1}^T Y_{t-1} + a_t y_t \right]$$

$$= \begin{bmatrix} \tilde{K}_t^{-1} & 0 \\ 0 & 1 \end{bmatrix} \left[P_{t-1} - q_t a_t^T P_{t-1} \right] \cdot \left[A_{t-1} Y_{t-1} + a_t y_t \right]$$

$$= \begin{bmatrix} \tilde{K}_t^{-1} & 0 \\ 0 & 1 \end{bmatrix} \left[P_{t-1} A_{t-1} Y_{t-1} \right] + \begin{bmatrix} \tilde{K}_t^{-1} & 0 \\ 0 & 1 \end{bmatrix}$$

$$\times \left[P_{t-1} a_t y_t - q_t a_t^T P_{t-1} A_{t-1} Y_{t-1} - q_t a_t^T P_{t-1} a_t y_t \right]$$

$$= \begin{bmatrix} \tilde{K}_t^{-1} & 0 \\ 0 & 1 \end{bmatrix} \left[P_{t-1} A_{t-1} Y_{t-1} \right] + \begin{bmatrix} \tilde{K}_t^{-1} & 0 \\ 0 & 1 \end{bmatrix}$$

$$\times \left[P_{t-1} a_t y_t - q_t a_t^T \left(\tilde{K}_{t-1} \right) \left(\tilde{K}_t^{-1} P_{t-1} A_{t-1} Y_{t-1} \right) - q_t a_t^T P_{t-1} a_t y_t \right]$$

(note: $\tilde{K}_{t-1}$ and $\tilde{K}_{t-1}^{-1}$ are $m \times m$)

$$= \begin{bmatrix} \tilde{K}_t^{-1} & 0 \\ 0 & 1 \end{bmatrix} \left[P_{t-1} A_{t-1} Y_{t-1} \right] + \begin{bmatrix} \tilde{K}_t^{-1} & 0 \\ 0 & 1 \end{bmatrix}$$

$$\times \left[P_{t-1} a_t y_t - q_t a_t^T \left(\tilde{K}_{t-1} \right) \left(\tilde{\alpha}_{t-1} \right) - q_t a_t^T P_{t-1} a_t y_t \right]$$

- As $\tilde{K}_t^{-1} P_{t-1} A_{t-1} Y_{t-1} = \tilde{\alpha}_{t-1}$:

$$\begin{bmatrix} \tilde{\alpha} \\ 0 \end{bmatrix} = \begin{bmatrix} \tilde{K}_t^{-1} & 0 \\ 0 & 1 \end{bmatrix} [P_{t-1}A_{t-1}Y_{t-1}] + \begin{bmatrix} \tilde{K}_t^{-1} & 0 \\ 0 & 1 \end{bmatrix}$$

$$\times \left[\left(P_{t-1} - q_t a_t^T P_{t-1} \right) a_t y_t - q_t a_t^T \tilde{K}_{t-1} \tilde{\alpha}_{t-1} \right]$$

$$= \begin{bmatrix} \tilde{K}_t^{-1} & 0 \\ 0 & 1 \end{bmatrix} [P_{t-1}A_{t-1}Y_{t-1}] + \begin{bmatrix} \tilde{K}_t^{-1} & 0 \\ 0 & 1 \end{bmatrix}$$

$$\times \left[\left(P_t \right) a_t y_t - q_t a_t^T \tilde{K}_{t-1} \tilde{\alpha}_{t-1} \right]$$

- As $P_t = P_{t-1} - q_t a_t^T P_{t-1}$

$$\begin{bmatrix} \tilde{\alpha} \\ 0 \end{bmatrix} = \begin{bmatrix} \tilde{K}_t^{-1} & 0 \\ 0 & 1 \end{bmatrix} [P_{t-1}A_{t-1}Y_{t-1}] + \begin{bmatrix} \tilde{K}_t^{-1} & 0 \\ 0 & 1 \end{bmatrix} \left[\left(P_t a_t y_t \right) - q_t a_t^T \tilde{K}_{t-1} \tilde{\alpha}_{t-1} \right]$$

$$= \begin{bmatrix} \tilde{K}_t^{-1} & 0 \\ 0 & 1 \end{bmatrix} [P_{t-1}A_{t-1}Y_{t-1}] + \begin{bmatrix} \tilde{K}_t^{-1} & 0 \\ 0 & 1 \end{bmatrix} \left[\left(q_t \right) y_t - q_t a_t^T \tilde{K}_{t-1} \tilde{\alpha}_{t-1} \right]$$

- As $P_t a_t = q_t$

$$\begin{bmatrix} \tilde{\alpha} \\ 0 \end{bmatrix} = \begin{bmatrix} \tilde{K}_t^{-1} & 0 \\ 0 & 1 \end{bmatrix} [P_{t-1}A_{t-1}Y_{t-1}] + \begin{bmatrix} \tilde{K}_t^{-1} & 0 \\ 0 & 1 \end{bmatrix} \cdot q_t \left[y_t - a_t^T \tilde{K}_{t-1} \tilde{\alpha}_{t-1} \right]$$

- Let $L_{t-1} = [P_{t-1}A_{t-1}Y_{t-1}] = \begin{bmatrix} L_{t-1}(1:m-1) \\ L_{t-1}(m) \end{bmatrix} \Rightarrow \begin{bmatrix} \tilde{K}_t^{-1} & 0 \\ 0 & 1 \end{bmatrix} [P_{t-1}A_{t-1}Y_{t-1}]$

$$= \begin{bmatrix} \tilde{K}_t^{-1} & 0 \\ 0 & 1 \end{bmatrix} \begin{bmatrix} L_{t-1}(1:m-1) \\ L_{t-1}(m) \end{bmatrix}$$

$$= \begin{bmatrix} \tilde{K}_t^{-1} \cdot L_{t-1}(1:m-1) \\ L_{t-1}(m) \end{bmatrix}$$

- Also, $\tilde{\alpha}_{t-1} = \tilde{K}_{t-1}^{-1} P_{t-1} A_{t-1} Y_{t-1}$

$$\Rightarrow \tilde{\alpha}_{t-1} = \begin{bmatrix} \tilde{K}_t^{-1} + \dfrac{\tilde{a}_s \tilde{a}_s^T}{\delta_s} & -\dfrac{a_s}{\delta_s} \\[2ex] -\dfrac{\tilde{a}_s^T}{\delta_s} & \dfrac{1}{\delta_s} \end{bmatrix} \begin{bmatrix} L_{t-1}(1:m-1) \\[1ex] L_{t-1}(m) \end{bmatrix}$$

$$\Rightarrow \begin{bmatrix} \tilde{K}_t^{-1} & 0 \\ 0 & 1 \end{bmatrix} [P_{t-1} A_{t-1} Y_{t-1}] = \begin{bmatrix} \tilde{K}_t^{-1} \cdot L_{t-1}(1:m-1) \\ L_{t-1}(m) \end{bmatrix}$$

$$= \begin{bmatrix} \tilde{\alpha}_{t-1}(1:m-1) + \tilde{a}_s \cdot \tilde{a}_{t-1}(m) \\ L_{t-1}(m) \end{bmatrix}$$

- So, $\begin{bmatrix} \tilde{\alpha} \\ 0 \end{bmatrix} = \begin{bmatrix} \tilde{\alpha}_{t-1}(1:m-1) + \tilde{a}_s \cdot \tilde{a}_{t-1}(m) \\ L_{t-1}(m) \end{bmatrix} + \begin{bmatrix} \tilde{K}_t^{-1} & 0 \\ 0 & 1 \end{bmatrix}$

- $\tilde{\alpha}_t = \tilde{\alpha}_{t-1}(1:m-1) + \tilde{a}_s \cdot \tilde{a}_{t-1}(m) + \tilde{K}_t^{-1} \cdot q_t(1:m-1) \cdot \left[y_t - a_t^T \tilde{K}_{t-1} \tilde{a}_{t-1} \right]$

7.3.9 Adding Element to Dictionary

■ $\mathcal{D}_t = \mathcal{D}_{t-1} + \{x_t\}$ and $m_t = m_{t-1} + 1$. $\tilde{K}_t$ grows accordingly, and the recursive formula for $\tilde{K}_t^{-1}$ is easily derived:

■ $\tilde{K}_t = \begin{bmatrix} \tilde{K}_{t-1} & \tilde{k}_{t-1}(x_t) \\ \tilde{k}_{t-1}(x_t)^T & k_{tt} \end{bmatrix}$

$$\tilde{K}_t^{-1} = \dfrac{1}{\delta_t} \begin{bmatrix} \delta_t \tilde{K}_{t-1}^{-1} + \tilde{a}_t \tilde{a}_t^T & -\tilde{a}_t \\ -\tilde{a}_t^T & 1 \end{bmatrix} \tag{7.15}$$

■ $A_t = \begin{bmatrix} A_{t-1} & 0 \\ 0^T & 1 \end{bmatrix}$

■ $A_t^T A_t = \begin{bmatrix} A_{t-1}^T A_{t-1} & 0 \\ 0^T & 1 \end{bmatrix}$

■ $P_t = (A_t^T A_t)^{-1} = \begin{bmatrix} (A_{t-1}^T A_{t-1})^{-1} & 0 \\ 0 & 1 \end{bmatrix}$

■ $P_t = \begin{bmatrix} P_{t-1} & 0 \\ 0^T & 1 \end{bmatrix}$

$$\tilde{\alpha}_t = \tilde{K}_t^{-1} P_t A_t^T Y_t$$

$$= \tilde{K}_t^{-1} \left(A_t^T A_t \right)^{-1} A_t^T Y_t$$

$$= \tilde{K}_t^{-1} \begin{bmatrix} \left(A_t^T A_t \right)^{-1} & 0 \\ 0 & 1 \end{bmatrix} \begin{bmatrix} A_{t-1} & 0 \\ 0 & 1 \end{bmatrix} \begin{bmatrix} Y_{t-1} \\ y_t \end{bmatrix} \tag{7.16}$$

$$= \tilde{K}_t^{-1} \begin{bmatrix} \left(A_t^T A_t \right)^{-1} A_{t-1} & 0 \\ 0 & 1 \end{bmatrix} \begin{bmatrix} Y_{t-1} \\ y_t \end{bmatrix}$$

$$= \tilde{K}_t^{-1} \begin{bmatrix} \left(A_t^T A_t \right)^{-1} A_{t-1} Y_{t-1} \\ y_t \end{bmatrix}$$

▪ Based on (7.11):

$$\tilde{\alpha}_t = \frac{1}{\delta_t} \begin{bmatrix} \delta_t \tilde{K}_{t-1}^{-1} + \tilde{a}_t \tilde{a}_t^T & -\tilde{a}_t \\ -\tilde{a}_t^T & 1 \end{bmatrix} \begin{bmatrix} P_{t-1} A_{t-1} Y_{t-1} \\ y_t \end{bmatrix}$$

$$= \frac{1}{\delta_t} \begin{bmatrix} \delta_t \tilde{K}_{t-1}^{-1} P_{t-1} A_{t-1} Y_{t-1} + \tilde{a}_t \tilde{a}_t^T P_{t-1} A_{t-1} Y_{t-1} - \tilde{a}_t y_t \\ -\tilde{a}_t^T P_{t-1} A_{t-1} Y_{t-1} + y_t \end{bmatrix}$$

$$= \begin{bmatrix} \left(\tilde{K}_{t-1}^{-1} P_{t-1} A_{t-1} Y_{t-1} \right) + \frac{1}{\delta_t} \tilde{a}_t \tilde{a}_t^T \cdot \tilde{K}_{t-1} \left(\tilde{K}_{t-1}^{-1} P_{t-1} A_{t-1} Y_{t-1} \right) - \frac{1}{\delta_t} \tilde{a}_t y_t \\ \frac{1}{\delta_t} y_t - \frac{1}{\delta_t} \tilde{a}_t^T \cdot \tilde{K}_{t-1} \left(\tilde{K}_{t-1}^{-1} P_{t-1} A_{t-1} Y_{t-1} \right) \end{bmatrix}$$

▪ Based on (7.16):

$$\tilde{\alpha}_{t-1} = \tilde{K}_{t-1}^{-1} P_{t-1} A_{t-1}^T Y_{t-1}$$

$$= \begin{bmatrix} \left(\tilde{\alpha}_{t-1} \right) + \frac{1}{\delta_t} \tilde{a}_t \left(\tilde{a}_t^T \tilde{K}_{t-1} \right) \left(\tilde{\alpha}_{t-1} \right) - \frac{1}{\delta_t} \tilde{a}_t y_t \\ \frac{1}{\delta_t} y_t - \frac{1}{\delta_t} \left(a_t^T \tilde{K}_{t-1} \right) \left(\tilde{\alpha}_{t-1} \right) \end{bmatrix}$$

■ We have $\tilde{a}_t = \tilde{K}_{t-1}^{-1} \cdot \tilde{k}_{t-1}(x_t)$, and we multiply each side with $\tilde{K}_{t-1}$ to obtain:

$$\tilde{K}_{t-1}\tilde{a}_t = \tilde{k}_{t-1}(x_t).$$

■ Then, replace in (7.16):

$$\tilde{\alpha}_{t-1} = \begin{bmatrix} (\tilde{\alpha}_{t-1}) + \dfrac{1}{\delta_t}\tilde{a}_t\left(\tilde{k}_{t-1}(x_t)^T\right)(\tilde{\alpha}_{t-1}) - \dfrac{1}{\delta_t}\tilde{a}_t y_t \\ \dfrac{1}{\delta_t}y_t - \dfrac{1}{\delta_t}\left(\tilde{k}_{t-1}(x_t)^T\right)(\tilde{\alpha}_{t-1}) \end{bmatrix} \tag{7.17}$$

■ Taking the transpose of each side of $\tilde{K}_{t-1}\tilde{a}_t = \tilde{k}_{t-1}(x_t)$

  – $\left(\tilde{K}_{t-1}\tilde{a}_t\right)^T = \tilde{k}_{t-1}(x_t)^T$

  – $\tilde{K}_{t-1}^T\tilde{a}_t^T = \tilde{k}_{t-1}(x_t)^T$

■ As $\tilde{K}_{t-1}^T = \tilde{K}_{t-1}$:

$$\tilde{k}_{t-1}(x_t)^T = \tilde{K}_{t-1}\tilde{a}_t^T \tag{7.18}$$

■ Using (7.18) in (7.17):

$$\tilde{\alpha}_t = \begin{bmatrix} (\tilde{\alpha}_{t-1}) + \dfrac{1}{\delta_t}\tilde{a}_t\left(y_t - \tilde{k}_{t-1}(x_t)^T \cdot \tilde{\alpha}_{t-1}\right) \\ \dfrac{1}{\delta_t}\left(y_t - \tilde{k}_{t-1}(x_t)^T \cdot \tilde{\alpha}_{t-1}\right) \end{bmatrix}$$

7.3.10 Dropping Element from Dictionary

■ Let us assume that element "*m*" was added at time *s*.
■ Also, assume that at time *t*, element "*m*" is dropped, then $\tilde{K}_t^{-1}$ should be equal to $\tilde{K}_{s-1}^{-1}$.

$$\tilde{K}_t^{-1} = \tilde{K}_{s-1}^{-1} \rightarrow \tilde{K}_{t-1}^{-1} = \tilde{K}_s^{-1} \tag{7.19}$$

■ Thus, $\tilde{K}_s^{-1} = \begin{bmatrix} \tilde{K}_{s-1}^{-1} + \dfrac{\tilde{a}_s\tilde{a}_s^T}{\delta_s} & -\dfrac{a_s}{\delta_s} \\ -\dfrac{\tilde{a}_s^T}{\delta_s} & \dfrac{1}{\delta_s} \end{bmatrix}.$

■ Replace $\tilde{K}_s^{-1}$ with $\tilde{K}_{t-1}^{-1}$ and $\tilde{K}_{s-1}^{-1}$ with $\tilde{K}_t^{-1}$ based on (7.19).

■ Subsequently,

$$\tilde{K}_{t-1}^{-1} = \begin{bmatrix} \tilde{K}_t^{-1} + \dfrac{\tilde{a}_s \tilde{a}_s^T}{\delta_s} & -\dfrac{\tilde{a}_s}{\delta_s} \\ -\dfrac{\tilde{a}_s^T}{\delta_s} & \dfrac{1}{\delta_s} \end{bmatrix} \tag{7.20}$$

■ $\tilde{K}_{t-1}^{-1}$ can also be written as

$$\begin{bmatrix} \tilde{K}_{t-1}^{-1}(1:m-1,1:m-1) & \tilde{K}_{t-1}^{-1}(1:m-1,m) \\ \tilde{K}_{t-1}^{-1}(m,1:m-1) & \tilde{K}_{t-1}^{-1}(m,m) \end{bmatrix} \tag{7.21}$$

$$= \begin{bmatrix} \tilde{K}_t^{-1} + \dfrac{\tilde{a}_s \tilde{a}_s^T}{\delta_s} & -\dfrac{\tilde{a}_s}{\delta_s} \\ -\dfrac{\tilde{a}_s^T}{\delta_s} & \dfrac{1}{\delta_s} \end{bmatrix}$$

■ $\tilde{K}_{t-1}^{-1}(1:m-1,1:m-1) = \tilde{K}_t^{-1} + \dfrac{\tilde{a}_s \tilde{a}_s^T}{\delta_s}$

$$\tilde{K}_t^{-1} = \tilde{K}_{t-1}^{-1}(1:m-1,1:m-1) - \dfrac{\tilde{a}_s \tilde{a}_s^T}{\delta_s} \tag{7.22}$$

■ $\tilde{K}_{t-1}^{-1}(m,m) = \dfrac{1}{\delta_s}$

$$\delta_s = \dfrac{1}{\tilde{K}_{t-1}^{-1}(m,m)} \tag{7.23}$$

■ $\tilde{K}_{t-1}^{-1}(1:m-1,m) = -\dfrac{\tilde{a}_s}{\delta_s}$

  – $\tilde{a}_s = -\tilde{K}_{t-1}^{-1}(1:m-1,m) \times \delta_s$, then based on (7.23)

$$\tilde{a}_s = \dfrac{-\tilde{K}_{t-1}^{-1}(1:m-1,m)}{\tilde{K}_{t-1}^{-1}(m,m)} \tag{7.24}$$

■ Example: Let

$$\tilde{K}_{10} = \begin{bmatrix} k_{11} & k_{12} \\ k_{21} & k_{22} \end{bmatrix}, \tilde{K}_{11} = \tilde{K}_{12} = \ldots = \tilde{K}_{19} = \tilde{K}_{10}. \tag{7.25}$$

■ Let $\tilde{K}_{10} = \tilde{K}_s = \begin{bmatrix} k_{11} & k_{12} & k_{13} \\ k_{21} & k_{22} & k_{23} \\ k_{31} & k_{32} & k_{33} \end{bmatrix} = \begin{bmatrix} \tilde{K}_{t-1} & \tilde{k}_{t-1}(x_t) \\ \tilde{k}_{t-1}(x_t)^T & k_{tt} \end{bmatrix}$

■ Let $s = 20$, then

$$\tilde{K}_{21} = \tilde{K}_{22} = \ldots = \tilde{K}_{29} = \tilde{K}_{20} \tag{7.26}$$

■ Let $t = 30$, then $\tilde{K}_t = \tilde{K}_{30}$ should equal $\tilde{K}_{10} = \tilde{K}_{19}$

■ Then, $\tilde{K}_s^{-1} = \tilde{K}_{20}^{-1} = \dfrac{1}{\delta_{20}} = \begin{bmatrix} \delta_{20}\tilde{K}_{19}^{-1} + \tilde{a}_{20}\tilde{a}_{20}^T & -\tilde{a}_{20} \\ -\tilde{a}_{20}^T & 1 \end{bmatrix}$, where $\tilde{a}_{20} = \tilde{K}_{19}^{-1} \times \tilde{k}_{19}(x_{20})$

$$\tilde{K}_{11}^{-1} = \ldots = \tilde{K}_{19}^{-1} = \tilde{K}_{10}^{-1} \tag{7.27}$$

$$\tilde{K}_{21}^{-1} = \ldots = \tilde{K}_{29}^{-1} = \tilde{K}_{20}^{-1} \tag{7.28}$$

■ Let $\tilde{K}_{29}^{-1} = \tilde{K}_{29}^{-1}(1:m-1,1:m-1)$
■ Then, comparing entities: $\tilde{K}_{29}^{-1}(1:m-1, 1:m-1) = \dfrac{1}{\delta_{20}} = \left[\delta_{20}\tilde{K}_{19}^{-1} + \tilde{a}_{20}\tilde{a}_{20}^T \right]$

  − $\delta_{20} \times \tilde{K}_{29}^{-1} = \left(\delta_{20}\tilde{K}_{19}^{-1} + \tilde{a}_{20}\tilde{a}_{20}^T \right)$
  − $\delta_{20}\tilde{K}_{29}^{-1} - \tilde{a}_{20}\tilde{a}_{20}^T = \delta_{20}\tilde{K}_{19}^{-1}$

$$\tilde{K}_{19}^{-1} = \frac{1}{\delta_{20}} \left[\delta_{20}\tilde{K}_{29}^{-1}(1:m-1,1:m-1) - \tilde{a}_{20}\tilde{a}_{20}^T \right] \tag{7.29}$$

■ As $\tilde{K}_{30} = \tilde{K}_{10}$ (before and after adding m) $\Rightarrow \tilde{K}_{30}^{-1} = \tilde{K}_{10}^{-1}$
■ Additionally, based on (7.27):
■ $\tilde{K}_{30}^{-1} = \tilde{K}_{19}^{-1}$
■ Using (29):

$$\tilde{K}_{30}^{-1} = \frac{1}{\delta_{20}} \left[\delta_{20}\tilde{K}_{29}^{-1}(1:m-1,1:m-1) - \tilde{a}_{20}\tilde{a}_{20}^T \right]$$

■ Replacing 30 with t and 20 with s:

$$\boxed{\tilde{K}_t^{-1} = \frac{1}{\delta_s} \left[\delta_s\tilde{K}_{t-1}^{-1}(1:m-1,1:m-1) - \tilde{a}_s\tilde{a}_s^T \right]}$$

7.3.10.1 Updating $\tilde{\alpha}_t$ after Dropping Element m

■ $\tilde{\alpha}_t = \tilde{\alpha}_t - \dfrac{1}{\delta_p} \begin{pmatrix} \tilde{a}_p \tilde{a}_p^T & -\tilde{a}_p \\ -\tilde{a}_p^T & 1 \end{pmatrix} \tilde{K}_t$

$\tilde{\alpha}_t = \tilde{K}_t^{-1} [P_t][A_t^T Y_t]$

$\qquad = \tilde{K}_t^{-1} \left[P_{t-1} A_{t-1}^T Y_{t-1} \right] + \tilde{K}_t^{-1} \left[P_{t-1} a_t y_t - q_t a_t^T \left(P_{t-1} A_{t-1}^T Y_{t-1} \right) - q_t a_t^T P_{t-1} a_t y_t \right]$

$\qquad = \tilde{K}_t^{-1} \left[P_{t-1} A_{t-1}^T Y_{t-1} \right]$

$\qquad\quad + \tilde{K}_t^{-1} \left[P_{t-1} a_t y_t - q_t a_t^T \left(\tilde{K}_{t-1} \right) \left(\tilde{K}_{t-1}^{-1} P_{t-1} A_{t-1}^T Y_{t-1} \right) - q_t a_t^T P_{t-1} a_t y_t \right]$

$\qquad = \tilde{K}_t^{-1} \left[P_{t-1} A_{t-1}^T Y_{t-1} \right] + \tilde{K}_t^{-1} \left[P_{t-1} a_t y_t - q_t a_t^T \left(\tilde{K}_{t-1} \right) (\tilde{\alpha}_{t-1}) - q_t a_t^T P_{t-1} a_t y_t \right]$

$$(7.30)$$

■ Based on (7.30):

$\tilde{\alpha}_t = \tilde{K}_t^{-1} \left[P_{t-1} A_{t-1}^T Y_{t-1} \right] + \tilde{K}_t^{-1} \left[P_{t-1} - q_t a_t^T P_{t-1} a_t y_t - q_t a_t^T \tilde{K}_{t-1} \tilde{\alpha}_{t-1} \right]$

$\qquad = \tilde{K}_t^{-1} \left[P_{t-1} A_{t-1}^T Y_{t-1} \right] + \tilde{K}_t^{-1} \left[(P_t) a_t y_t - q_t a_t^T \tilde{K}_{t-1} \tilde{\alpha}_{t-1} \right]$

$\qquad = \tilde{K}_t^{-1} \left[P_{t-1} A_{t-1}^T Y_{t-1} \right] + \tilde{K}_t^{-1} \left[(P_t a_t) y_t - q_t a_t^T \tilde{K}_{t-1} \tilde{\alpha}_{t-1} \right]$

$\qquad = \tilde{K}_t^{-1} \left[P_{t-1} A_{t-1}^T Y_{t-1} \right] + \tilde{K}_t^{-1} \left[(q_t) y_t - q_t a_t^T \tilde{K}_{t-1} \tilde{\alpha}_{t-1} \right]$

$\qquad = \tilde{K}_t^{-1} \left[P_{t-1} A_{t-1}^T Y_{t-1} \right] + \tilde{K}_t^{-1} q_t \left[y_t - a_t^T \tilde{K}_{t-1} \tilde{\alpha}_{t-1} \right]$

■ Replacing $\tilde{K}_t^{-1}$ based on (7.29)

$\tilde{\alpha}_t = \left[\tilde{K}_{t-1}^{-1}(1:m-1,1:m-1) - \dfrac{\tilde{a}_s \tilde{a}_s^T}{\delta_s} \right] \left[P_{t-1} A_{t-1}^T Y_{t-1} \right] + \tilde{K}_t^{-1} q_t \left[y_t - a_t^T \tilde{K}_{t-1} \tilde{\alpha}_{t-1} \right]$

$\qquad = \left[\tilde{K}_{t-1}^{-1}(1:m-1,1:m-1) \left[P_{t-1} A_{t-1}^T Y_{t-1} \right] - \dfrac{\tilde{a}_s \tilde{a}_s^T}{\delta_s} P_{t-1} A_{t-1}^T Y_{t-1} \right.$

$\qquad\quad \left. + \tilde{K}_t^{-1} q_t \left[y_t - a_t^T \tilde{K}_{t-1} \tilde{\alpha}_{t-1} \right] \right]$

■ Assuming $\left[\tilde{K}_{t-1}^{-1}(1:m-1,1:m-1) \cdot P_{t-1} A_{t-1}^T Y_{t-1} \right] \approx \tilde{\alpha}_{t-1}(1:m-1)$

we get: $\tilde{\alpha}_t \approx \left[\tilde{\alpha}_{t-1}(1:m-1)\right] - \dfrac{\tilde{a}_s \tilde{a}_s^T}{\delta_s} P_{t-1} A_{t-1}^T Y_{t-1} + \tilde{K}_t^{-1} q_t \left[y_t - a_t^T \tilde{K}_{t-1} \tilde{\alpha}_{t-1}\right]$

- where the approximation for $\tilde{\alpha}_{t-1}(1:m-1)$ is valid up to additive term in the $(m-1)$ components of $\tilde{\alpha}_{t-1}$.

- $\tilde{\alpha}_t = \tilde{\alpha}_{t-1}(1:m-1) - \dfrac{\tilde{a}_s \tilde{a}_s^T}{\delta_s}\left(\tilde{K}_{t-1}\right)\left(\tilde{K}_t^{-1} P_{t-1} A_{t-1}^T Y_{t-1}\right) + \tilde{K}_t^{-1} q_t \left[y_t - a_t^T \tilde{K}_{t-1} \tilde{\alpha}_{t-1}\right]$

- Keep in mind that $\tilde{K}_{t-1}$ and $\tilde{K}_{t-1}^{-1}$ are full $m \times m$ matrices.

- $\tilde{\alpha}_t = \tilde{\alpha}_{t-1}(1:m-1) - \dfrac{\tilde{a}_s \tilde{a}_s^T}{\delta_s}\left(\tilde{K}_{t-1}\right)\left(\tilde{\alpha}_{t-1}\right) + \tilde{K}_t^{-1} q_t \left[y_t - a_t^T \tilde{K}_{t-1} \tilde{\alpha}_{t-1}\right]$

- Here, the size of $\left[a_t^T \tilde{K}_{t-1} \tilde{\alpha}_{t-1}\right]$ is $[(1 \times m)(m \times m)(m \times 1)] = [(1 \times m)(m \times 1)] = (1 \times 1)$

- The size of $\left[y_t - a_t^T \tilde{K}_{t-1} \tilde{\alpha}_{t-1}\right]$ is (1×1) and the size of $[q_t]$ is (max_$m \times 1$), where max_m denotes the maximum value that $m(t)$ ever reached.

- The size of $\tilde{K}_{t-1}^{-1}$ is $(m-1) \times (m-1)$.

- The size of $\dfrac{\tilde{a}_s \tilde{a}_s^T}{\delta_s}$ is $[((m-1) \times 1)(1 \times (m-1))] = (m-1) \times (m-1)$.

- The size of $\left[\tilde{K}_{t-1} \tilde{\alpha}_{t-1}\right]$ is $[(m \times m)(m \times 1)] = (m \times 1)$.

- Thus, approximate as follows:

$$\tilde{\alpha}_t \approx \tilde{\alpha}_{t-1}(1:m-1) - \dfrac{\tilde{a}_s \tilde{a}_s^T}{\delta_s}\left[\tilde{K}_{t-1} \tilde{\alpha}_{t-1}\right]_{1:m-1} + \tilde{K}_t^{-1} q_t(1:m-1)\left[y_t - a_t^T \tilde{K}_{t-1} \tilde{\alpha}_{t-1}\right].$$

- Once again approximation is valid up to some additive term to entities in $\tilde{\alpha}_t$.

- Then, size of $[\tilde{\alpha}_t] = (m-1) \times 1$.

7.4 Discussion

Security engineering is defined as building systems to remain dependable in the face of malice, error, or mischance [27]. A comprehensive security engineering solution should encompass multiple cyclic processes, including business process review, risk analysis, vulnerability analysis, engineered solutions, testing, and documentation [28]. There are various engineering solutions to predict, prevent, detect, and respond security breaches ranging from hardware-level solutions to the knowledge of applied psychology. Today's organizations used a wide variety of tools such as classical signature-based antiviruses, rule-based firewalls, IDSs and IPSs, and statistical, machine learning, and deep learning-based anomaly detection engines. Mathematical theories are the backbone of the statistical, machine learning, and deep learning methods; as a result, they play a significant role in developing powerful security engineering solutions. Kernel online anomaly detection algorithm is one of the acceptable applications in the field of incremental anomaly detection in the academia. We believe that it also has an intrinsic potential to be improved and applied to different industry sectors.

7.5 Conclusions

This chapter provides the overview of assets, threat, vulnerability, attack, risk, and their relationships. It emphasizes on the points that differentiate information security from security engineering. This chapter also emphasizes the importance of defining, predicting, and preventing attack factors. It presents the current state of various anomaly detection methods in the literature and elaborates on the effectiveness of machine learning techniques for detecting network anomalies and attacks. The deficiency of offline network anomaly detection algorithms constitutes an important point to start for addressing research and development in the field of online network anomaly detection algorithms. We provided the in-depth review of the KOAD algorithm. The step-by-step mathematical proofs of the algorithm were presented. We also emphasized on the low computational cost (memory and time) of the algorithm by presenting the mathematical background of algorithm. We believe that this study can be used as a guideline for researcher and practitioners in the field of network anomaly detection. The algorithm should be improved from two points of view: First, setting parameters and thresholds of the algorithm in an automated way; second, upgrading the algorithm to solve classification problems. Attempts in these directions were made in [29,30], but they still remain work-in-progress.

Acknowledgments

This project was initiated when Tarem Ahmed was a Visiting Scholar in the laboratory of Prof. Aydin Alatan at the Department of Electrical and Electronics Engineering, Middle East Technical University, Ankara, Turkey, on an Erasmus+ KA107 International Credit Mobility Program with the Department of Electrical and Electronic Engineering, BRAC University, Dhaka, Bangladesh.

Notes

* A matrix differentiation operator is defined as $\dfrac{\partial}{\partial A} \triangleq \begin{bmatrix} \dfrac{\partial}{\partial a_{11}} & \cdots & \dfrac{\partial}{\partial a_{1n}} \\ \vdots & \ddots & \vdots \\ \dfrac{\partial}{\partial a_{m1}} & \cdots & \dfrac{\partial}{\partial a_{1m}} \end{bmatrix}.$

† For finding the inverse of non-square matrix, the pseudo-inverse matrix is used. If the columns of a matrix A are linearly independent, then we should calculate the pseudo-inverse with $A^+ = (A^T \cdots A)^{-1} \cdots A^T$. However, if the rows of the matrix are linearly independent, we should calculate the pseudo-inverse with $A^+ = A^T (A^T \cdots A)^{-1}$.

‡ Matrix inversion lemma assumes that $M = \begin{bmatrix} A & B \\ C & D \end{bmatrix}$ is an investable matrix and made of invertible blocks such as A, B, C, and D. Subsequently, prove that

$$\begin{bmatrix} A & B \\ C & D \end{bmatrix}^{-1} = (A - B \cdot D^{-1} \cdot C)^{-1} = A^{-1} + A^{-1} \cdot B \cdot (D - C \cdot A^{-1} \cdot B)^{-1} \cdot C \cdot A^{-1}.$$

A and BCD have the same dimensions. It is linear algebra trick, which is applicable in kernel theory [26].

References

1. A.-S. K. Pathan, R. A. Saeed, M. A. Feki, and N. H. Tran, "Guest editorial: Special issue on integration of IoT with future internet," *Journal of Internet Technology*, vol. 15, no. 2, pp. 145–147, 2014.

2. "Industrial IoT adoption worldwide as of 2017, by region." www.statista.com/statistics/797390/industrial-iot-adoption-worldwide-by-region/, 2017. [Online; accessed 01-Feb-2019].

3. M. Ahmed, A. N. Mahmood, and J. Hu, "A survey of network anomaly detection techniques," *Journal of Network and Computer Applications*, vol. 60, pp. 19–31, 2016.

4. T. Ahmed, R. Rahman, and A. Pathan, "Survey of anomaly detection algorithms: Towards self-learning networks," in *Security of Self-Organizing Networks: MANET, WSN, WMN, VANET*, pp. 65–89, Auerbach Publications, CRC Press, Jeju Island, Korea, 2010.

5. A. Lazarevic, L. Ertoz, V. Kumar, A. Ozgur, and J. Srivastava, "A comparative study of anomaly detection schemes in network intrusion detection," in *Proceedings of the 2003 SIAM International Conference on Data Mining*, pp. 25–36, SIAM, San Francisco, CA, May 1–3 2003.

6. V. Hodge and J. Austin, "A survey of outlier detection methodologies," *Artificial Intelligence Review*, vol. 22, no. 2, pp. 85–126, 2004.

7. M. Markou and S. Singh, "Novelty detection: A review part 1: Statistical approaches," *Signal Processing*, vol. 83, no. 12, pp. 2481–2497, 2003.

8. M. Markou and S. Singh, "Novelty detection: A review part 2: Neural network based approaches," *Signal Processing*, vol. 83, no. 12, pp. 2499–2521, 2003.

9. A.-S. K. Pathan, *The State of the Art in Intrusion Prevention and Setection*, Auerbach Publications, New York, 2014.

10. R. Reid and J. Van Niekerk, "From information security to cyber security cultures," in *2014 Information Security for South Africa*, pp. 1–7, IEEE, Johannesburg, South Africa, 2014.

11. "Information technology security techniques code of practice for information security management," Standard, International Organization for Standardization, 2005.

12. M. S. Merkow and J. Breithaupt, *Information Security: Principles and Practices*, Pearson Education, Indianapolis, IN, 2014.
13. M. H. Bhuyan, D. K. Bhattacharyya, and J. K. Kalita, "Network anomaly detection: methods, systems and tools," *IEEE Communications Surveys & Tutorials*, vol. 16, no. 1, pp. 303–336, 2013.
14. M. Mohammed and A.-S. K. Pathan, *Automatic Defense Against Zero-Day Polymorphic Worms in Communication Networks*, Auerbach Publications, New York, 2016.
15. M. Ahmed and A.-S. K. Pathan, "Investigating deep learning for collective anomaly detection-an experimental study," in *International Symposium on Security in Computing and Communication*, pp. 211–219, Springer, Bangalore, India, September 19–22 2018.
16. T. Ahmed, M. Coates, and A. Lakhina, "Multivariate online anomaly detection using kernel recursive least squares," in INFOCOM 2007. *26th IEEE International Conference on Computer Communications*. IEEE, pp. 625–633, IEEE, Anchorage, AK, May 2007.
17. T. Ahmed, B. Oreshkin, and M. Coates, "Machine learning approaches to network anomaly detection," in *Proceedings of the 2nd USENIX Workshop on Tackling Computer Systems Problems with Machine Learning Techniques*, pp. 1–6, USENIX Association, Cambridge, MA, April 10 2007.
18. B. Scholkopf and A. J. Smola, *Learning with Kernels: Support Vector Machines, Regularization, Optimization, and Beyond*, MIT press, Cambridge, MA, 2001.
19. Y. Engel, S. Mannor, and R. Meir, "The kernel recursive least-squares algorithm," *IEEE Transactions on Signal Processing*, vol. 52, no. 8, pp. 2275–2285, 2004.
20. T. Ahmed, S. Ahmed, and F. E. Chowdhury, "Taking meredith out of grey's anatomy: Automating hospital ICU emergency signaling," in *2016 IEEE International Conference on Acoustics, Speech and Signal Processing (ICASSP)*, pp. 1886–1890, IEEE, Shanghai, China, March 2016.
21. A. Anika, K. L. Karim, R. Muntaha, F. Shahrear, S. Ahmed, and T. Ahmed, "Multi image retrieval for kernel-based automated intruder detection," in *2017 IEEE Region 10 Symposium (TENSYMP)*, pp. 1–5, IEEE, Kochi, India, 2017.
22. T. Ahmed, S. Ahmed, S. Ahmed, and M. Motiwala, "Real-time intruder detection in surveillance networks using adaptive kernel methods," in *2010 IEEE International Conference on Communications*, pp. 1–5, IEEE, Delhi, India, 2010.
23. T. Ahmed, X. Wei, S. Ahmed, and A.-S. K. Pathan, "Efficient and effective automated surveillance agents using kernel tricks," *Simulation*, vol. 89, no. 5, pp. 562–577, 2013.
24. T. Ahmed, A.-S. K. Pathan, and S. S. Ahmed, "Learning algorithms for anomaly detection from images," in *Biometrics: Concepts, Methodologies, Tools, and Applications*, pp. 281–308, IGI Global, Hershey, PA, 2017.
25. T. Ahmed, A.-S. K. Pathan, and S. Ahmed, "Adaptive algorithms for automated intruder detection in surveillance networks," in *2014 International Conference on Advances in Computing, Communications and Informatics (ICACCI)*, pp. 2775–2780, IEEE, Delhi, India, September 2014.
26. G. Strang, *Introduction to Linear Algebra*, vol. 3, Wellesley-Cambridge Press, Wellesley, MA, 1993.
27. R. Anderson, *Security Engineering: A Guide to Building Dependable Distributed Systems*, John Wiley & Sons, Chichester, UK, 2008.
28. R. B. Vaughn, "A practical approach to sufficient infosec," in *National Information System Security Conference*, pp. 1–11, Citeseer, Baltimore, MD, 1996.

29. T. Ahmed, S. Ahmed, and A.-S. K. Pathan, "Automated surveillance in distributed visual networks: An empirical comparison of recent algorithms," *International Journal of Control and Automation*, vol. 7, no. 3, pp. 389–400, 2014.

30. H. Islam and T. Ahmed, "Anomaly clustering based on correspondence analysis," in *2018 IEEE 32nd International Conference on Advanced Information Networking and Applications (AINA)*, pp. 1019–1025, IEEE, Krakow, Poland, May 2018.

31. B. Haasdonk and H. Burkhardt, "Invariant kernel functions for pattern analysis and machine learning," *Machine Learning*, vol. 68, no. 1, pp. 35–61, July 2007. https://link.springer.com/article/10.1007/s10994-007-5009-7

Chapter 8

Secure Addressing Protocols for Mobile Ad Hoc Networks

Uttam Ghosh
Vanderbilt University

Pushpita Chatterjee
Old Dominion University

Raja Datta
IIT Kharagpur

Al-Sakib Khan Pathan
Southeast University, Dhaka, Bangladesh

Danda B. Rawat
Howard University

Contents

8.1 Introduction

Internet of Things (IoT) is a heterogeneous network of physical devices and other objects embedded with sensors and actuators. It enables the objects to connect and communicate a large amount of data to offer a new class of advanced services available at anytime, anywhere, and for anyone. IoT consists of various types of wireless networks such as wireless sensor and ad hoc networks (WiFi, ZigBee, and RFID) to make the physical infrastructures such as buildings (homes, schools, offices, factories, etc.), utility networks (electricity, gas, water, etc.), transportation networks (roads, railways, airports, harbors, etc.), transportation vehicles (cars, rails, planes, etc.), healthcare systems, and information technology networks smarter, secure, reliable, and fully automated. It collects, stores, and communicates a large volume of heterogeneous data from various types of networks and provides critical services in manufacturing, healthcare, utility, and transportation networks.

A *Mobile Ad hoc Network* (MANET) is a collection of devices equipped with wireless communications and networking capability [1]. These nodes can be arbitrarily located and are free to move randomly at any given time. Node mobility can vary from almost stationary to constantly moving nodes. Network topology and interconnections between the nodes can change rapidly and unpredictably. As MANET is infrastructure-less and highly dynamic in nature, unique node identification is a very important requirement in such setting. This is required for a node to participate in unicast communications and routing, and also to retain its identity when a network gets partitioned and/or merged due to its dynamicity.

Originally, MANET was conceived as a small isolated ad hoc network that does not have any connection to the outer world. However, over the years, MANET has evolved into an important network having applications in various fields that requires communication with other infrastructure networks. This has necessitated MANET's connectivity with the Internet, albeit in a restricted manner. This means that usually, MANET will work in isolation but occasionally may connect to an infrastructure network (e.g., Internet or IoT) if need arises. This brings us to an important issue of node identification in accordance with the usual IP network. Moreover, IP address facilitates multi-hop routing in the network and also when the network is connected with IoT [2].

Dynamic Host Configuration Protocol (DHCP) [3] provides static or dynamic address allocation to the network nodes that can be manual or automatic. Manual or static address configuration in most cases is inapplicable as the nodes in MANET are highly mobile leading to partitioning/merging of networks. Further, the centralized DHCP is not a suitable solution, as it has to maintain configuration information of all the nodes in the network.

The address allocation protocol needs to consider several challenges and dynamic scenarios as MANET is a distributed and dynamic network. In the simplest scenario, a node can join and leave the network at any time as shown in Figure 8.1. A new node *N* needs an IP address whenever it joins the network. An allocated IP address can be reused when the node departs from the network (the node either switches off itself or leaves the network due to node mobility).

In MANET, the mobile nodes are free to move arbitrarily and one or more configured nodes go out of others' transmission ranges for a while. As a result of this, the network may get partitioned as shown in Figure 8.2a. These partitions may grow independently and continue communications with their IP addresses. Due to node mobility, these partitions may merge later. If a new node *N* joins a partition and obtains an IP address belonging to the other partition, then conflict occurs when these two partitions merge as presented in Figure 8.2b.

Figure 8.3 shows another scenario where two separately configured MANETs merge. There may be two different cases. In the first case (for instance), there is no address conflict as both MANET 1 and MANET 2 have different network identifiers. In the second case, there are some duplicate addresses as the address allocation

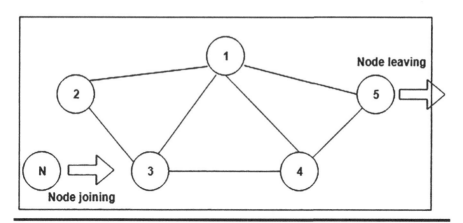

Figure 8.1   Node joining and leaving in MANET.

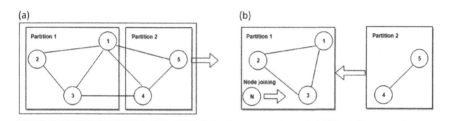

Figure 8.2   (a) Network partitions; (b) network merging.

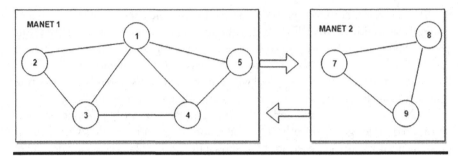

Figure 8.3 Network merges of two independent MANETs.

in MANET 1 is independent of MANET 2. As a result of this, some (or all) nodes in one MANET may need to change their addresses.

It can be seen from the above discussion that there is a need to design a secure distributed addressing protocol for MANETs that can provide compatible connectivity with IoT. Thus, for assigning addresses in MANETs, a standard addressing protocol should have the following objectives [2,4]:

- **Distributed Dynamic Address Configuration**: As MANET is infrastructure-less, and its nodes are mobile in nature, the addressing protocol for providing dynamic IP addresses to the new nodes should be a distributed one (rather than centralized). If the addressing protocol is centralized, then it has to rely heavily on duplicate address detection (DAD) mechanism for address resolution in the network. This causes broadcast storm problem.
- **Uniqueness**: The protocol should assign unique IP addresses to the nodes of the network for correct routing and for point-to-point communication.
- **Robustness**: The chances of address conflicts among the nodes due to network partitioning and network merging should be as low as possible.
- **Scalability**: As the network grows, the time taken to obtain an IP address (i.e., *addressing latency*) and the number of message exchanges (i.e., *addressing overhead*) during address allocation should be minimum.
- **Security**: The protocol should be able to withstand attacks while allocating addresses to the nodes of a MANET. The major security threats associated with dynamic IP address configuration in MANET are as follows:
  - **Address spoofing attack**: In this attack, an IP address of a node can be spoofed by a malicious host to hijack the network traffic.
  - **Address exhaustion attack**: Here, an attacker may claim as many IP addresses as possible for exhausting all the valid IP addresses so as to prevent a newly arrived node from getting an IP address.
  - **False address conflict attack**: In this attack, an attacker may transmit address conflict messages falsely so that a victim node may give up its current address and seek for a new one.

- **False deny message attack**: Here, an attacker may continuously transmit false deny messages to prevent a newly arrived node from getting an IP address.
- **Reply attack**: In this type of attack, an attacker injects previously captured packets (from an authorized node) into the network. This attack can also be seen at other layers of the protocol stacks.

■ **Organization of the chapter**: This chapter presents the main challenges related to addressing protocols in dynamic and ad hoc environment like MANET. It further categorizes the existing addressing protocols and provides a detailed overview of them in Section 8.2. In Section 8.3, the chapter also evaluates and compares the performance of addressing protocols of MANET through a set of performance metrics. Finally, it concludes with future research direction related to addressing in MANET.

8.2 Address Allocation Protocols for MANET

This section presents a brief review on recently proposed dynamic address allocation protocols [2,5,6,7–29] for MANET to enable proper communication in the network. In order to adapt to the dynamic environment of a MANET, these protocols bear many similarities to each other, such as self-organizing, self-healing behavior. However, these approaches also differ in a wide range of aspects, such as address format, address allocation information, usage of centralized servers or full decentralization, hierarchical structure or flat network organization, and explicit or implicit DAD mechanism. According to [30], all the existing IP address allocation schemes for ad hoc networks can be classified into *stateless allocation* and *stateful allocation* approaches.

8.2.1 Stateless Allocation Approaches

In *stateless* allocation approaches, nodes in a network do not store any address allocation information. These approaches are mainly based on self-configuration. Each node chooses its address randomly and then performs DAD to ensure uniqueness of the chosen address within the network. The major disadvantages of stateless approaches are high addressing overhead and latency.

Most of the existing address allocation algorithms for a MANET use DAD mechanism [5] to resolve address conflict in the network. Figure 8.4 shows the DAD process. DAD is required whenever a new node joins a MANET or independent networks merge. A new node picks up a tentative IP address and uses DAD process to determine whether this address is available or not. All the nodes having a valid IP address participate in DAD mechanism to protect their IP addresses being used by a new node. In order to detect the uniqueness of the address, the node sends a *Duplicate Address Probe* (DAP) message and expects an

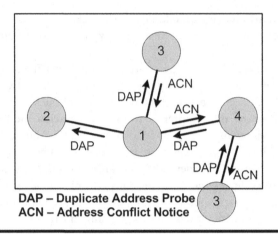

Figure 8.4 DAD mechanism.

Address Conflict Notice (ACN) message back in a certain timeout period. If no ACN message is received, the node may assume that the address is not in use.

In [5], the authors propose a mechanism called weak DAD. The basic idea of weak DAD is that the duplicate addresses may be tolerated as long as packets reach the destination node intended by the sender even if the destination node's address is also being used by another node. Here, each node has a unique key which is included in the routing packets and in the routing table entries. Even if the IP address conflicts exist, they can be identified by their unique keys. The main drawback of weak DAD is its dependency on the routing protocol and also traffic overhead caused by the integration of the key value in routing packets.

In *passive DAD* [6], nodes use periodic link state routing information to notify other nodes about their neighbors. This is usually very costly and will result in serious redundancy, contention, and collision, which is called *broadcast storm* problem [31].

The *ad hoc address autoconfiguration* (AAA) scheme [7] under the stateless allocation approaches uses randomly selected addresses from the address range of 169.254/16. Thereafter, to guarantee the uniqueness of the selected address, DAD is performed by each node of the network. During this process, an *Address Request* message is flooded in the network to query the usage of its tentative address. If the address is already in use, an *Address Reply* message is unicast back to the requesting node. The requesting node then re-initiates the address allocation process. However, this approach does not consider complex scenarios such as network partitions and merger.

Fazio, Villari, and Puliafito have proposed an *automatic IP address configuration* (AIPAC) scheme [8] for MANET. In the AIPAC address allocation scheme, there are two phases. In the initialization phase, two distinct nodes enter each other's range by using the Host Identifier (HID). The AIPAC then locates the nodes without a valid IP address using the HID. The node that has the higher HID then selects the configuration parameters for both the interacting nodes. When a new node (Requester) wants to join the network, it receives the IP address through one of the configured nodes (Initiator). The Initiator negotiates for the Requester's valid IP address in the allocation phase, corrects the configuration, and then offers it to the Requester.

In [9], Wang, Reeves, and Ning have proposed a secure auto-configuration scheme that uses self-authentication technique. By using one-way hash function, it binds a node's address with a public key. Address owner can use the corresponding public key to unilaterally authenticate itself. In this scheme, whenever a node, P, wants to join the network, it first randomly generates a public/private key pair and a secret key. Thereafter, node P calculates a 32-bit (in IPv4) or a 128-bit (in IPv6) hash value of the public key, that is, $IP = H(public\ key)$, where H is a secure one-way hash function. Next, node P temporarily uses this IP address, initiates a timer, and broadcasts a DAP message throughout the network to verify the uniqueness of the IP address. If a configured node (call it Q) finds that the IP address in a received DAP message is same as its own, it verifies the authenticity of this DAP message. If the verification result is positive, Q then broadcasts an ACN message to inform the corresponding node P of the address conflict. Otherwise, node Q simply discards the received DAP message. The scheme handles network partitioning/merging by employing the concept of passive DAD mechanism.

Passive autoconfiguration for mobile ad hoc networks (PACMAN) [10] has been proposed by Weniger where a new node selects an address using a probabilistic algorithm. In order to verify the uniqueness of the addresses, it makes use of passive DAD in conjunction with a distributed maintenance of a common allocation table. When it detects that two nodes are using the same IP address, it reports the problem to one of them using a unicast message to change its address. Moreover, it takes into account the problem of changing an address that has some ongoing communication. To fix this, while changing an address, a node notifies other nodes (with whom communication is going on) of its new IP address, so that they can make an encapsulation of the messages properly.

8.2.2 Stateful Allocation Approaches

In *stateful* allocation approaches, the nodes in the network keep track of assigned and free addresses for address assignment as well as network management. Each node in *MANETconf* scheme [11], presented by Nesargi and Prakash under the stateful allocation approach, maintains a list of IP addresses which are in use in the network. A new node X obtains an address through an existing node Y in the MANET; thereafter, the node Y performs an address query throughout the MANET. This address allocation requires a modified DAD for checking address duplication. Here, a positive acknowledgment (ACK) is required from all the known nodes indicating that the IP address is available for use. This may result in an ACK explosion. Network partitions and mergers are detected throughout the modified DAD.

In [12], Zhou, Ni, and Mutka propose a scheme called Prophet for MANET. The scheme proposes a function $f(n)$ to generate a series of random numbers for address allocation. The desired properties of function $f(n)$ are as follows:

i. A sequence satisfies the extremely long interval between two occurrences of the same number.
ii. The probability that the function returns the same number for two different state values is very low.

The protocol works as follows: The first node M in the MANET generates a random number and sets its IP address. It also uses a random state value as a seed for its function $f(n)$. Another node N can get its IP address from node M along with a state value as a seed for its $f(n)$. Whenever a node joins the network, same process continues for address allocation of new nodes.

Figure 8.5 shows an example of Prophet address allocation where $f(n) =$ (address*state*11) mod 17 and the effective address range is [1, 16]. The first node M randomly generates 5 as its address and the seed. When N enters, node M changes its state of $f(n)$ to 3 (=5*5*11 mod 17) and assigns it to N as address and seed. Node O and node P get addresses 12 (=3*5*11 mod 17) and 14 (=3*3*11 mod 17), respectively. The main advantages of this protocol are its low addressing latency and low communication overhead. However, its major drawback is that even with a large address space, address conflicts may exist in the network. To resolve these address conflicts, it requires mechanisms such as passive DAD or weak DAD.

In [13], a *Dynamic Address Configuration Protocol* named DACP has been presented for MANET. In DACP, an elected address authority maintains the state information of MANET. In the initialization phase, a node chooses two addresses: *temporary* and *tentative* addresses. The temporary address is used to verify the uniqueness of the tentative address using DAD. The tentative address is made

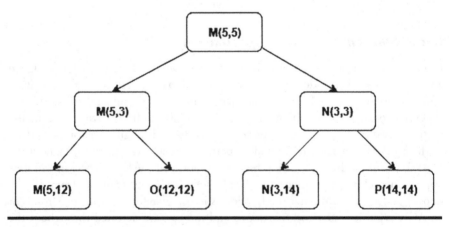

Figure 8.5 An example of Prophet address allocation.

permanent after it is found that the same is not a duplicate address. After successful address configuration is made, a new node registers its tentative address with the address authority to make it permanent. Here, the new node waits for an advertisement from the address authority for a certain period of time. After receiving the said advertisement, the new node sends a registration request and waits for the registration confirmation message from the address authority. The concerned new node can use this address only after this confirmation message is received. The overhead due to duplication detection mechanism and the high periodic flooding of the messages due to registration with address authority are the major drawbacks of this protocol.

Another protocol used for dynamic address allocation is the optimized DACP [14], in which, instead of discovering the server, the server itself periodically broadcasts the address request to reduce the overhead of broadcasting. However, the latency for the hosts to obtain the addresses in this scheme is much higher.

Taghiloo et al. have presented an address auto-configuration based on *virtual address space mapping* (VASM) scheme [15]. According to this protocol, when a new node wants to join the network, it sends a single-hop message in order to find an *Initiator*. If no reply message is received for this packet, it assumes that it is the only node in that network and starts the network setup process. If the node receives more than one response, it selects the sender of the first arrived packet as *Initiator* and sends it an address request packet. The main task of Initiator is to obtain a new IP address from its *Allocator* and assign it to the requesting node (*Requester*). In order to balance the overhead of protocol traffic and minimize the addressing latency, an *Allocator* can create another *Allocator* in the network for generating unique addresses for the new nodes. For an efficient management of network events (such as merging and partitioning), each *Allocator* holds a list of all *Allocators* in the network. As the number of *Allocators* in each network is limited, the size of *Allocators'* list will be very small. Tajamolian, Taghiloo, and Tajamolian have proposed a lightweight secure address configuration scheme [16]. For address allocation, VASM address configuration scheme is used [15]. To secure the address allocation, the scheme uses a zero knowledge approach.

In [17], X. Chu et al. proposed a quadratic residue-based address allocation scheme for mobile ad hoc networks. In this scheme, when a new node wants to join a MANET, it sends a DISCOVER message to obtain an IP address. If no reply is received, it assumes that it is the first node in the network. The node configures itself with an IP address and also generates an address block. Thereafter, a new node can obtain an address and also an address block from a configured node. In this way, network grows up from one to many in the network. To handle network merging and partitioning, this scheme uses DAD to remove duplicate addresses from the network.

In the *Buddy* system allocation scheme [18], each node maintains a block of free addresses. A configured node which receives an *Address Request* from a new node assigns the requesting node an IP address from its block of free addresses. It also divides its block of free addresses into two equal parts and gives one half to the requesting node and the other half it keeps with itself for future use. However, it is always difficult for the individual nodes to manage such type of address blocks in a MANET and is also complex to implement. Cavalli and Orset presented a secure hosts auto-configuration scheme [19]. The scheme employs the concept of challenge, where a node has to answer a question to prove its identity. Here, a new node X sends a request with its public key and a temporary identifier to its neighbors. The neighbors then calculate a nonce, cipher it with the public key of the node X, and then return it to the node X. The new node X, after having deciphered it with its private key, returns the nonce to the concerned nodes. It uses the *Buddy* system technique to allocate the IP addresses.

Another dynamic address configuration scheme called Prime DHCP [20] is proposed by Hsu and Tseng. In this scheme, address can be allocated to the new host without broadcasting it over the whole of MANET. Prime DHCP makes each host a DHCP proxy and runs a prime numbering address allocation (*PNAA*) algorithm individually to compute unique addresses for address allocation. According to *PNAA* algorithm, the first node that creates the network is called the root node. The *PNAA* algorithm works based on the following two rules:

i. The root node having address 1 can allocate prime numbers in an ascending order.
ii. Other nodes (not the root) can allocate addresses equal to its own address multiplied by the unused prime number, starting from the largest prime factor of its own address.

Figure 8.6 shows an example of *PNAA* address allocation tree. For example, let us take the node having address 9. As the largest prime factor of 9 is 3, it can allocate the sequence of addresses $9*3, 9*5, 9*7$, and so on up to the largest address bounded by address space. Therefore, it eliminates the need of DAD mechanism. Prime DHCP [20] uses Destination-Sequenced Distance-Vector (DSDV) [32] routing protocol in order to handle network partitions and mergers.

MMIP [21], ADIP [22], IDDIP [23], IDSDDIP [24], and SD-RAC [2] are the address allocation schemes proposed by Ghosh and Datta, where the nodes of the network act as proxies and are able to allocate addresses independently to the new nodes. None of these schemes needs to run DAD mechanism to verify the uniqueness of addresses in the network. In order to provide authentication, MMIP binds the hardware address with the IP address at the time of address allocation. The authentication for address configuration is done with the help of a trusted third party in case of ADIP scheme, whereas IDDIP scheme uses self-authentication technique. IDSDDIP [24] provides security using a RSA-based cryptography system, and SD-RAC [2] uses a bilinear pairing-based signature scheme.

Figure 8.7 partially shows an example of how unique IPv6 address can be allocated by a node acting as proxy in SD-RAC [2]. For simplicity, they present the

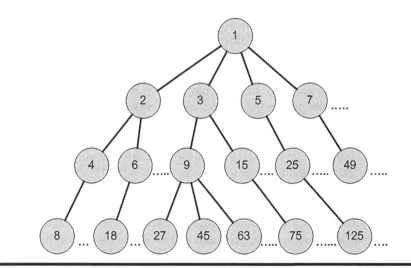

Figure 8.6   An example of *PNAA* address allocation tree.

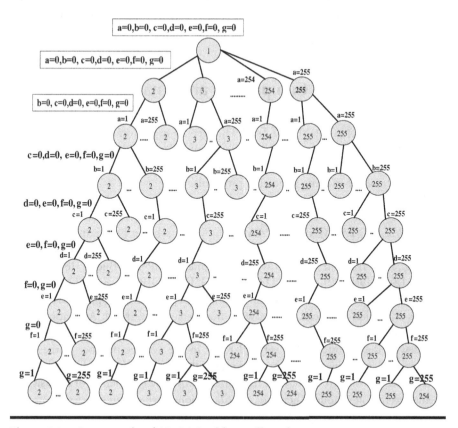

Figure 8.7   An example of SD-RAC address allocation.

addresses in dotted decimal format, that is, m.n.o.p.q.r.s.t.a.b.c.d.e.f.g.h wherein m.n.o.p.q.r.s.t is the network prefix (fixed for a network) and a.b.c.d.e.f.g.h is a host identifier (a, b, c, d, e, f, and g are variables). In the figure, the last byte (h) of an IP address is shown within the circle and the other bytes (a, b, c, d, e, f, and g) are shown outside the circle. The root proxy has an IP m.n.o.p.q.r.s.t.0.0.0.0.0.0.0.1, and the host identifier addressing it can assign a range from 1.0.0.0.0.0.0.1 to 255.0.0.0.0.0.0.1 and from 0.0.0.0.0.0.0.2 to 0.0.0.0.0.0.0.255 with network prefix m.n.o.p.q.r.s.t. A proxy having host identifier 1.0.0.0.0.0.0.1 can allocate addresses from 1.1.0.0.0.0.0.1 to 1.255.0.0.0.0.0.1. This scheme can allocate addresses from m.n.o.p.q.r.s.t.0.0.0.0.0.0.0.1 to m.n.o.p.q.r.s.t.255.255.255.255.255.255.255.254 in the network uniquely.

A *filter-based addressing protocol* (FAP) [25] has been proposed by Fernandes et al., which maintains a distributed database stored in filters containing currently allocated addresses in a compact fashion. Bloom filter is used to detect and resolve network partitioning and merging, and sequence filter is used to obtain IP addresses. Then, a node acquires an IP address based on sequence filter, and it needs to run a DAD in the whole network to update sequence filters of all other nodes residing in the network. Here too, we find that the addressing overhead and latency would be high due to DAD process running in the entire network.

8.3 Performance Comparison and Analysis

Table 8.1 presents the qualitative comparison of the existing dynamic address allocation approaches for MANETs [2,33]. Here, n is the total number of mobile nodes, l is the average number of links between nodes, d is the network diameter, and t is the average 1-hop latency. The following metrics are used to analyze the performance of the existing allocation protocols.

Uniqueness: One of the most important metrics in address allocation technique is to guarantee the uniqueness of the allocated addresses. Stateless protocols (i.e., AIPAC [8], PACMAN [10], AAA [7], Wangs scheme [9]) cannot guarantee the uniqueness of allocated addresses. On the contrary, stateful protocols can do that (e.g., MANETconf [11], ODACP [14], Prime DHCP [20], Buddy [18], Cavallis scheme [19], ADIP [22], IDDIP [23], FAP [25], IDSDDIP [24], and SD-RAC [2]).

Addressing Overhead: Addressing overhead of an addressing protocol includes address allocation overhead and network maintenance overhead. Allocation overhead refers to the average number of messages required for an address allocation to a new node. Maintenance overhead refers to the average number of messages received by a node per time slot to maintain the network. In stateless protocols, DAD needs to run throughout the network to verify the uniqueness of an assigned IP address to a new node. Therefore, the allocation overhead for stateless protocols due to flooding is $O(n^2)$. Stateless protocols do not rely on the underlying routing protocols and prefer to use the reactive routing protocol (e.g., AODV [26]) for the

Table 8.1 Performance Comparison of Address Allocation Protocols

Protocol	IP Family	Allocation Approach	Addressing Latency	Overhead		Complexity	Robustness	Scalability	Security
				Allocation	Maintenance				
MANET CONF	IPV4	Stateful	$O(2td)$	$O(n^2)$	$O(2l/n)$	High	Low	Low	No
FAP	IPV4	Stateful	$O(2td)$	$O(n^2)$	$O(2l/n)$	High	Low	Low	No
AIPAC	IPV4	Stateless	$O(2td)$	$O(n^2)$	$O(2l/n)$	Medium	Low	Low	No
PACMAN	IPV4, IPV6	Stateless	$O(2td)$	$O(n^2)$	$O(2l/n)$	Medium	Low	Low	No
AAA	IPV4	Stateful	$O(2td)$	$O(n^2)$	$O(2l/n)$	Medium	Low	Low	No
ODACP	IPV4	Stateful	$O(2td)$	$O(2l)$	$O(2l)$	Low	Low	Low	No
PROPHET	IPV4	Stateful	$O(2t)$	$O(2l/n)$	$O(2l/n)$	High	Low	Medium	No
PRIME DHCP	IPV4	Stateful	$O(2t)$	$O(2l/n)$	$O(2l/n)$	Low	Medium	Medium	No
BUDDY	IPV4, V6	Stateful	$O(2t)$	$O(2l/n)$	$O(n^2)$	High	Medium	Low	No
CAVALI [19]	IPV4, IPV6	Stateful	$O(2t)$	$O(2l/n)$	$O(n^2)$	High	Medium	Low	Yes

(Continued)

Table 8.1 (Continued) Performance Comparison of Address Allocation Protocols

Protocol	IP Family	Allocation Approach	Addressing Latency	Overhead		Complexity	Robustness	Scalability	Security
				Allocation	Maintenance				
WANG [9]	IPV4, IPV6	Stateless	$O(2t)$	$O(n^2)$	$O(2l/n)$	Medium	Low	Low	Yes
ADIP	IPV4	Stateful	$O(2t)$	$O(2l/n)$	$O(2l/n)$	Low	Medium	Medium	Yes
IDDIP	IPV4	Stateful	$O(2t)$	$O(2l/n)$	$O(2l/n)$	Low	Medium	Medium	Yes
IDSDDIP	IPV6	Stateful	$O(2t)$	$O(2l/n)$	$O(n^2)$	Low	Medium	High	Yes
SDRAC	IPV6	Stateful	$O(2t)$	$O(2l/n)$	$O(2l/n)$	Low	High	High	Yes

network maintenance. The maintenance overhead of stateless allocation protocols is $O(2l/n)$ as the reactive routing protocol needs to exchange a one-hop broadcast message to track the neighbors periodically.

On the other hand, most of the stateful address allocation protocols need to send a one-hop broadcast message to get an address from the network. Therefore, allocation overhead for these protocols is $O(2l/n)$. However, the allocation overhead of MANETcon [11] and FAP [25] is $O(n^2)$. This is because both of the protocols require a positive ACK from all known nodes indicating that the address is available for use. The allocation overhead of ODACP [14] is $O(2l)$ as each node has to register with an address authority. Stateful allocation protocols use either a reactive routing protocol or a proactive routing protocol for the network maintenance. Therefore, if the addressing protocol uses the proactive routing protocol, then the network maintenance overhead is $O(n^2)$; otherwise, the overhead is $O(2l/n)$.

Addressing Latency: Addressing latency is the time between points when a node requests for an address and when it acquires the address from the network. DAD mechanism is used by stateless allocation protocols where an *Address Request* message is flooded in the whole network. A timer for Address Reply is set based on the network diameter, and a new node can configure itself when the timer times out. Therefore, the addressing latency of the stateless protocols is $O(2td)$.

In stateful allocation, most of the protocols send a one-hop broadcast message for an address allocation; hence, their addressing latency is the round-trip delay time, that is, $O(2t)$. However, the addressing latency of FAP [25], ODACP [14], and MANETconf protocols is $O(2td)$. The addressing protocols that consider the security aspects have an additional latency.

Complexity: The addressing protocol should use the network resources (e.g., power and memory of nodes, bandwidth of the network) as minimal as possible during address allocation. The complexity of the stateless allocation protocols is considered to be medium as they generate addresses from some random numbers. Under the stateful address allocation, FAP [25], MANETconf [11], Prophet [12], Cavalli et al. [19], and Buddy [18] protocols are highly complex to implement and synchronize the addresses among the nodes in the network. In ODACP [14], each node needs to register with the centralized leader node and eliminates the need of flooding the messages, hence reducing the message complexity. Other protocols under the stateful category have low computational complexity as these protocols do not require maintaining the address blocks and complex functions to generate addresses. These protocols generate addresses by the existing network nodes acting as proxies. This reduces the complexity and memory requirements of these protocols even further.

Robustness: Robustness refers to the adaptability of an addressing protocol in a dynamic network environment, including network partitioning and merging. The addressing protocol is considered to be highly robust if it can guarantee the address uniqueness even when the network gets partitioned and merges again. Here, two scenarios need to be considered when a new node joins the network. In the first scenario, a network grows independently and never partitions into multiple networks or merges with the other partitions or networks. In the second scenario, a network starts independently, and subsequently, the network partitions into multiple networks due to node mobility. Further, new nodes can join any partition and these partitions may merge with each other. In stateless allocation, Prophet and FAP, the addresses are generated randomly; hence, there is a chance of address conflict in both of the above scenarios. Therefore, the robustness of these protocols is considered to be low. On the other hand, most of the stateful allocation protocols ensure address uniqueness in the first scenario, but the address conflict may exist in the second scenario. Thus, the robustness of these protocols is considered to be medium. SD-RAC [2] protocol ensures address uniqueness in both scenarios. Therefore, the protocol is considered to be highly robust.

Scalability: The scalability of an addressing protocol is said to be high if the performance of the protocol does not degrade much in terms of addressing latency and addressing overhead even when the network size is large. The addressing overhead and the addressing latency of the stateless allocation protocols and MANETconf [11], FAP [25], and ODACP [14] protocols under the stateless allocation category are $O(n^2)$ and $O(2td)$, respectively. Therefore, these protocols are considered to be of low scalability. In stateful allocation, if a new node can acquire its address from a neighbor node and the addressing belongs to IPv4 family, then most of the nodes acting as proxies may exhaust their address spaces very quickly in a large network. This increases the addressing overhead and the addressing latency of the protocols. The scalability of these protocols is medium. On the contrary, if a new node can acquire its address from a neighbor node and the addressing belongs to IPv6 family, then the nodes acting as proxies have larger address spaces. Therefore, these protocols are highly scalable.

Security: The addressing protocol must provide security against the attacker and ensure that only the authorized nodes are configured and granted access to the network resources. Under the stateless allocation, only the protocol [9] proposed by Wang et al. considers the security aspects. In stateful allocation, ADIP [22], IDDIP [23], IDSDDIP [24], SD-RAC [2], and the protocol [19] proposed by Cavalli et al. consider the security at the time of address allocation process. Table 8.2 presents the secure address allocation protocols for MANETs.

As discussed earlier, MANET can be an isolated network or can be connected with Internet (and also with IoT) using gateway nodes [34–37]. The above address allocation protocols can allocate addresses to the nodes locally. This network uses these allocated addresses for unicast communication within the network. However, it uses network address translator (NAT) to translate the local address to the global address whenever the network needs to connect with IoT. The gateway node runs the NAT protocol for the connectivity between MANET and IoT.

Table 8.2   Secure Address Allocation Protocols

| Protocols | Attacks | | | | | Security Mechanisms |
	Address Spoofing	Address Exhaustion	Address Conflict	False Deny	Negative Reply	
CAVALI [19]	Protected	Protected	Protected	Protected	Protected	Challenge response
WANG [9]	Protected	Protected	Not protected	Protected	Protected	Self-authentication
ADIP	Protected	Not protected	Protected	Protected	Not protected	Trusted third party
IDDIP	Protected	Not protected	Protected	Protected	Protected	Self-authentication
IDSDDIP	Protected	Not protected	Protected	Protected	Protected	Trusted third party
SDRAC	Protected	Protected	Protected	Protected	Protected	Self-authentication

8.4 Conclusions

A number of address allocation protocols for mobile ad hoc networks have been presented in this chapter. Here, the existing allocation protocols have been categorized into *stateful* and *stateless*. Their performances are shown in terms of addressing overhead, latency, robustness, scalability, and security. It can be seen that most of the address allocation protocols do not consider the security aspect. As a result of this trend, there would be several types of potential attacks during the phase of address allocation to the nodes. These protocols also assume that the gateway node provides the connectivity between the MANET and IoT using NAT. However, there will be always several challenges to connect the MANET with IoT. Efficient mechanisms need to be devised to do the needful using optimal resources and cost given the configuration of future network settings.

References

1. P. Chatterjee, U. Ghosh, I. Sengupta, and S. K. Ghosh, "A trust enhanced secure clustering framework for wireless ad hoc networks," *Wireless Networks*, vol. 20, no. 7, pp. 1669–1684, 2014.
2. U. Ghosh and R. Datta, "A secure addressing scheme for large-scale managed MANETs," *IEEE Transactions on Network and Service Management*, vol. 12, no. 3, pp. 483–495, 2015.
3. R. Droms, "Dynamic host configuration protocol," *RFC 2131*, March 1997.
4. U. Ghosh and R. Datta. "A novel signature scheme to secure distributed dynamic address configuration protocol in mobile ad hoc networks," in *Proceedings 2012 IEEE Wireless Communications and Networking Conference (WCNC)*, (Paris, France), pp. 2700–2705, 2012.
5. N. H. Vaidya, "Weak duplicate address detection in mobile ad hoc networks," in *Proceedings of ACM International Symposium on Mobile Ad Hoc Networking and Computing (MobiHoc02)*, (Lausanne, Switzerland), pp. 206–216, June 2002.
6. K. Weniger, "Passive duplicate address detection in mobile ad hoc networks," in *Proceedings of IEEE WCNC*, (Florence, Italy), February 2003.
7. C. E. Perkins, J. T. Malinen, R. Wakikawa, E. M. Belding-Royer, and Y. Sun, "Ad hoc address autoconfiguration," *IETF Internet Draft draft-ietf-manet-autoconf-01.txt*, November 2001.
8. M. Fazio, M. Villari, and A. Puliafito, "Aipac: Automatic ip address configuration in mobile ad hoc networks," *Computer Communications*, vol. 29, no. 8, pp. 1189–1200, 2006.
9. P. Wang, D. S. Reeves, and P. Ning, "Secure address auto-configuration for mobile ad hoc networks," in *Proceedings of 2nd Annual International Conference MobiQuitous*, (San Diego, CA), pp. 519–522, 2005.
10. K. Weniger, "Pacman: Passive autoconfiguration for mobile ad hoc networks," *Special issue, IEEE JSAC, Wireless Ad Hoc Networks*, vol. 23, pp. 507–519, March 2005.

11. S. Nesargi and R. Prakash, "Manetconf: Configuration of hosts in a mobile ad hoc network," in *Proceedings of IEEE INFOCOM*, (New York), pp. 1059–1068, 2002.
12. H. Zhou, L. M. Ni, and M. W. Mutka, "Prophet address allocation for large scale MANETs," in *Proceedings of IEEE INFOCOM*, (San Francisco, CA), pp. 1304–1311, 2003.
13. Y. Sun and E. M. Belding-Royer, "Dynamic address configuration in mobile ad hoc networks," *UCSB Technical Report*, pp. 2003–2011, June 2003.
14. Y. Sun and E. M. Belding-Royer, "A study of dynamic addressing techniques in mobile ad hoc networks," *Wireless Communications and Mobile Computing*, vol. 4, pp. 315–329. doi:10.1002/wcm.215, April 2004.
15. M. Taghiloo, M. Dehghan, J. Taghiloo, and M. Fazio, "New approach for address auto-configuration in manet based on virtual address space mapping (vasm)," in *Proceedings of IEEE ICTTA 2008*, (Damascus, Syria), 7–11 April 2008.
16. M. Tajamolian, M. Taghiloo, and M. Tajamolian, "Lightweight secure ip address auto-configuration based on vasm," in *Proceedings of 2009 International Conference on Advanced Information Networking and Applications Workshops*, (Bradford, UK), pp. 176–180, 2009.
17. X. Chu, Y. Sun, K. Xu, Z. Sakander, and J. Liu, "Quadratic residue based address allocation for mobile ad hoc networks," in *IEEE International Conference on Communications (ICC)*, vol. 158, (Beijing, China), pp. 2343–2347, 19–23 May 2008.
18. M. Mohsin and R. Prakash, "Ip address assignment in a mobile ad hoc network," in *Proceedings of IEEE MILCOM*, September 2002.
19. A. Cavalli and J. Orset, "Secure hosts auto-configuration in mobile ad hoc networks," *Elsevier Ad Hoc Networks*, vol. 3, no. 5, pp. 656–667, 2005.
20. Y. Hsu and C. Tseng, "Prime dhcp: A prime numbering address allocation mechanism for manets," *IEEE Communications Letters*, vol. 9, no. 8, August 2005.
21. U. Ghosh and R. Datta, "Mmip: A new dynamic ip configuration scheme with mac address mapping for mobile ad hoc networks," in *Proceedings of Fifteenth National Conference on Communications 2009*, (IIT Guwahati, India), January 2009.
22. U. Ghosh and R. Datta, "Adip: An improved authenticated dynamic ip configuration scheme for mobile ad hoc networks," *International Journal of Ultra Wideband Communications and Systems*, vol. 1, pp. 102–117, 2009.
23. U. Ghosh and R. Datta, "A secure dynamic ip configuration scheme for mobile ad hoc networks," *Elseiver Ad Hoc Networks*, vol. 9, no. 7, pp. 1327–1342, 2011.
24. U. Ghosh and R. Datta. "IDSDDIP: A secure distributed dynamic IP configuration scheme for mobile ad hoc networks," *International Journal of Network Management*, vol. 23, no. 6, pp. 424–446, 2013.
25. N. C. Fernandes, M. D. D. Moreira, and O. C. M. B. Duarte, "An efficient and robust addressing protocol for node autoconfiguration in ad hoc networks," *IEEE/ACM Transactions on Networking*, vol. 21, no. 99, p. 1, 2013.
26. C. Perkins, E. Belding-Royer, and S. Das, "Ad hoc on-demand distance vector (aodv) routing," *draft-ietf-manet-aodv-11.txt*, June 2002 (work in progress).
27. M. Taghiloo, J. Taghiloo, and M. Dehghan, "A survey of secure address autoconfiguration in MANET," in *Proceedings of 10th IEEE International Conference on Communication Systems (ICCS)*, pp. 1–5, October 2006.

28. W. Xiaonan and Q. Huanyan, "Cluster-based and distributed ipv6 address configuration scheme for a MANET," *Wireless Personal Communications*, vol. 71, no. 4, pp. 3131–3156, doi:10.1007/s11277-013-0995-1, August 2013.

29. T. R. Reshmi and K. Murugan, "Secure and Reliable Autoconfiguration Protocol (SRACP) for MANETs," *Wireless Personal Communication*, vol. 89, no. 4, pp. 1243–1264, doi:10.1007/s11277-016-3314-9, August 2016.

30. N. Wangi, R. Prasad, M. Jacobsson, and I. Niemegeers, "Address autoconfiguration in wireless ad hoc networks: protocols and techniques," *IEEE Wireless Communications*, vol. 15, pp. 70–80, February 2008.

31. S. Ni, Y. Tseng, Y. Chen, and J. Sheu, "The broadcast storm problem in a mobile ad hoc network," in *Proceedings of the ACM/IEEE MOBICOM*, (Seattle, WA), pp. 151–162, 1999.

32. C. E. Perkins and P. Bhagwat, "Highly dynamic destination-sequenced distance vector routing (dsdv) for mobile computers," *SIGCOMM Compututer Communication Review*, vol. 24, no. 4, pp. 234–244, 1994.

33. S. Kim and J. Chung, "Message complexity analysis of mobile ad hoc network address autoconfiguration protocols," *IEEE Transactions on Mobile Computing*, vol. 7, no. 3, pp. 358–371, March 2008.

34. R. Kushwah, S. Tapaswi, and A. Kumar, "A detailed study on internet connectivity schemes for mobile ad hoc network," *Wireless Personal Communication*, vol. 104, no. 4, pp. 1433–1471, doi:10.1007/s11277-018-6093-7, February 2019.

35. C. E. Perkins, J. T. Malinen, R. Wakikawa, A. Nilsson, and A. J. Tuominen, "Internet connectivity for mobile ad hoc networks," *Wireless Communications and Mobile Computing*, vol. 2, no. 5, 465–482, 2002.

36. P. Bellavista, G. Cardone, A. Corradi, and L. Foschini, "Convergence of manet and wsn in iot urban scenarios," *IEEE Sensors Journal*, vol. 13, pp. 3558–3567, October 2013.

37. B. KameswaraRao and A. S. N. Chakravarthy, "A taxonomical review on manet networks for iot based air pollution controls," *International Journal of Computer Science and Technology*, vol. 8, no. 3, pp. 44–49, July–September 2017.

CYBER CHALLENGES

Chapter 9

Cryptographic Attacks, Impacts and Countermeasures

Hazaa Al Fahdi and Mohiuddin Ahmed

Edith Cowan University

Contents

9.1 Introduction

Cryptography is the use of mathematics to produce encrypted data and to decrypt it. Cryptography involves two main operations: encryption and decryption. Both of the operations ensure information security, which includes confidentiality, integrity, and availability. Moreover, most commercial companies, governments, and other business fields are using cryptography to ensure data security. For example, banks use an encryption algorithm to encrypt payments and money transactions of their clients. Thus, the user of the encryption algorithms will ensure the banks' services security and will avoid any breaches of the data [1]. Furthermore, most encryption algorithms have been attacked by different cyber-attacks. For example, the data encryption standard (DES) is symmetric that is used to encrypt electronic data. DES has a key length of 56 bits, and due to its key length, it has been tracked several times by brute force attack [1].

Due to the increase in the usage of encryption, everyone is aiming to secure his/her data. This led to increasing cyber security attacks against cryptography algorithms. The attacks left many impacts and risks behind them on their users such as governments, business, and normal users. The risk is high to loss sensitive data, damage systems, or financial losses. Therefore, the users of encryption features are required to implement a set of countermeasures to ensure that the usage of the encryption algorithm is secure, and also all business data or secret operations data are safe while performing encryption processes and transferring of the data. Otherwise, hackers can breach their data due to the lack of security. Moreover, different types of software, hardware, and policies can be implemented as best practice solutions to prevent different types of attacks against cryptography algorithms [1].

9.1.1 Chapter Roadmap

This chapter is divided into four main sections, which are as follows: Section 9.1 is about introduction of the chapter. Section 9.2 is about an overview of cryptography concept, Section 9.3 is focusing on different types of cryptography attacks, and Section 9.4 is about the impacts of cryptography attacks, and the last section is about countermeasures of the cryptography attacks.

9.2 Cryptography Overview

Cryptography is a method of altering and transferring sensitive data in an encryption process so that only who has an authorization can reveal it and work on it. Cryptography is a Greek word in which "crypto" means secret or hidden and "graphy" means writing [1]. Thus, cryptography is the process of hiding the original data to produce another data that is hard to understand, which ensure confidentiality, non-repudiation, integrity, and authenticity for the transferred data [2]. The cryptography has different number of components used in both encryption and decryption processes. Basically, any cryptographic system will have the following components:

A. Plain text: It is the data which is to be secured while communication [2].
B. Ciphertext: It is the changed plain text which cannot be read or understand when looking at it. To read it, it is required to apply the decryption algorithm. The ciphertext might or might not be secure, which depends on the encryption algorithm [2].
C. Encryption algorithm: It is the way to decode the plain text to produce unreadable text. This can be used in specific algorithms such as advanced encryption standard (AES) or DES. Also, it is used at the sender's side [2].
D. Decryption algorithm: It is the opposite process of the used encryption algorithm. Thus, it uses ciphertext and decryption key to reveal the original plain text [2].
E. Encryption key: This key is used with the encryption algorithm to produce the ciphertext of the plain text. The key is either known only to the sender or the receiver usually, and it depends on the algorithm used [2].
F. Decryption key: This key is used within the decryption algorithm in order to obtain the plain text from the ciphertext. The key is either known only to the sender or the receiver usually, and it depends on the algorithm used [2].

9.2.1 Cryptography Algorithms

Modern cryptography is based on strong scientific methods. The new cryptography is divided into several types. In this book, this chapter will discuss symmetric key and asymmetric key cryptography algorithms. An encryption method is based on one key of encryption and decryption processes. Both the receiver and the sender use the same key. This method was the only one that was known for encryption until June 1961. The symmetric cryptography of encryption and decryption is based on the study of the block ciphers and stream ciphers [1].

A block cipher is a modern embodiment of Alberti's polyalphabetic cipher. The block cipher takes a block of plain text as an input and a key; this block will go through an encryption process to produce an output of block of ciphertext in the same size [3]. On the other hand, a stream cipher is opposite to the block cipher. The stream cipher is taking the key and combined with the plain text as bit-by-bit or character-by-character. The output of stream cipher is based on altering the plain text orders to generate the stream cipher [3].

9.2.1.1 Symmetric Key Cryptography

Symmetric encryption is one kind of encryption method that uses one key for both encryption and decryption operations. The key which is used for encryption and decryption is called secret key, and it is only accessible by a specific user. Moreover, the symmetric encryption uses two types of encryption algorithms. The first algorithm is called block algorithm, which has different lengths of bits that encrypted in different blocks by the secret key. The encrypted data will be held until all the blocks complete. The second is the stream algorithm in which all data are encrypted as a stream rather than reserved in the memory [4]. Furthermore, symmetric method is fast for encryption and decryption operations compared to asymmetric method. Also, it has several types of algorithms such as AES, DES, RC4, RC5, and IDEA encryption algorithms. Moreover, symmetric is used to encrypt a large value of data such as database encryption of an organization, payment for banking, and online store [5].

Figure 9.1 shows the encryption and decryption processes where the plain text is encrypted and decrypted by using one key for the two parties.

The DES and the AES are block ciphers based on symmetric-key cryptography. Both the algorithms have different features, weakness, and strengths. The DES encrypts a block of 64 bits by using a key of 56 bits in length. The encryption process includes 16 rounds of confusion and diffusion. The confusion is an encryption operation that makes the relation between the key and ciphertext obscured.

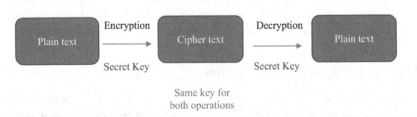

Figure 9.1 Symmetric encryption.

The diffusion is also an encryption operation that gathers one of the plain text values over many ciphertext values to produce a strong ciphertext. In addition, it is not recommended to use the DES due to its key length as well as other aspects of the algorithm that have been kept secret by the US government [1].

Furthermore, the AES can encrypt a block of 128 bits. The AES has different sizes of keys of 128, 192, and 256 bits. Each key length uses a different number of subkeys. Moreover, the AES is more secured compared with DES, because it uses large size of keys, which makes it strong against some types of cyber security attacks such as brute force attack. Also, the AES is part of known standards such as IPsec and TLS, which make the users more confident to implement the AES in their software, hardware, communications, and other types of field that required to make the data secured and encrypted [1].

9.2.1.2 Asymmetric Key Cryptography

The asymmetric cryptosystem also called public-key cryptosystem has two different types of keys, namely, the public key and the private key. The public key is usually used for encryption by the sender, where the plain text is combined with the key to produce the ciphertext (depends on the algorithm used); also, the public key can be distributed publicly. The private key is used by the receiver for decryption operation, and it must be kept secret [1]. In addition, several protocols used asymmetric cryptography, for example, SSH, SSL, OpenPGP, and digital signature. Also, Rivest–Shamir–Adleman (RSA) is using asymmetric method, which provides security for different channels. For example, RSA is using SSL/TLS protocol to provide secure communication of users over a network. Moreover, asymmetric method provides digital signature, which ensures the identification of a sender or receiver of a document [1].

Figure 9.2 shows the encryption and decryption processes in asymmetric cryptography. The encryption process is using the public key, and the private key is used to decrypt the encrypted data.

9.2.2 The Purpose of Cryptography

The usage of cryptography is to ensure confidentiality, integrity, non-repudiation, and authentication. These features are divided into cryptography algorithms. For example, the symmetric cryptography such as AES can ensure the confidentiality

Figure 9.2 Asymmetric cryptography.

and authentication of a communication between two parties by key session. Also, the non-repudiation and authentication can be found while using the RSA and DSA of asymmetric cryptography algorithms [1]. In spite of this, the cryptography is a useful technique to secure communication and data transmission, but there are several breaches on cryptography methods that led to many impacts on the usage of the cryptography and its users [5].

9.3 Cryptography Attacks

In this day, the usage of information is not limited to technology devices, but to all human life aspects such as communication, for example, emails, fax, and social media applications. The share of critical information has become an important thing to ensure the confidentiality and integrity of transmitted data. Thus, most of the organizations, business, and governments give high attention to their sensitive information by paying a high amount to implementing policies, software, and hardware to make sure their data is safe. Unfortunately, some of the cryptography algorithms have failed to secure the encrypted data against several attacks [6,7].

The attackers can apply passive or active attacks on a target to get some information about it. The passive attack is playing a sensitive role to obtain information on a system or user device without authorization to access. Also, the passive attack does not affect the systems resources, but affects the data's confidentiality, for example, eavesdropping on two parties' communication channel. This is done without the two parties knowing which usually leads to stealing the information [7].

On the other hand, the active attack aims to modify or damage a system's resources or operation in a way to obtain the wanted data. The active attack leaves high effects on the target systems and threats the integrity and availability of it [1].

Table 9.1 reflects several attacks that have been approved against the cryptography algorithms.

To discuss in depth about each of the above attacks, the following sections explain more about them.

9.3.1 Brute Force Attack

A brute force attack is one of the techniques used to reveal an important data for a specific target such as password of a system, encryption keys of different types of encryption algorithms, and SSH logins. According to researchers, the brute force is used against encryption algorithm. For example, the brute force attack has been approved against the DES due to its key length of 56 bits. Thus, the attacker can apply the attack to obtain encrypted data within a few hours [5]. Furthermore, the brute force attack depends on a specific wordlist or an intelligent database, which is used to detect the algorithm key or password of the target. In spite of brute force

Table 9.1 Attacks Key Features

Attack Name	Key Feature(s)
Brute force attack	Reveals sensitive information such as password, encryption, and decryption keys and SSH logins
Dictionary attack	Reveals passwords, algorithms keys
Rainbow table	Cracks passwords or hashes
John the Ripper	Crack passwords through a wordlist Runs in different operation systems
Side-channel attack	Gathers vulnerabilities of a system or cryptography algorithm Gains access to encrypted data
Probing attacks	Scans a system, networks, algorithms to discover vulnerabilities
Implementation attack	Exploits the vulnerability of an application, software, or algorithms
Fault analysis	Gathers a high amount of erroneous information to gain access to sensitive information
Chosen-ciphertext attack	Based on selecting numbers of ciphertext to obtain decryption data by using a random or unknown key
COA	Uses only a set of ciphertext to decrypt it
PAA	Determines the secret key by gathering specific information about cryptography algorithms
Birthday attack	Discloses a sensitive information
KPA	Aims to use the known plain text to decrypt the set of ciphertext
CPA	Determines the encryption key of cryptography algorithms through choosing a plain text

useful attack, it is limited to short keys. For example, the AES has different key lengths of 128, 192, or 256 bits. Therefore, the brute force attack would take a long time to attack 128 bits of AES key [8].

Furthermore, there are several tools or techniques used by brute force attacks such as dictionary attack, Rainbow table, and John the Ripper.

Dictionary attack: The attackers use a dictionary attack to obtain sensitive information of an encryption algorithm such as decryption key. The process is to build a dictionary of ciphertexts, keys, and identical plain text, and then to try all possible possibilities in a period of time. The dictionary attack can be a success due to a lack of security of users, for example, usage of poor password and sharing secret pair keys with other parties. Thus, the attackers can take advantage of users' faults and build the dictionary to reveal the sensitive information of the target [9].

Rainbow table: This attack is usually used to crack passwords or hashes that have been created of plain text or password. The cracking method is to try each possible hash to obtain encrypted data [10].

John the Ripper: A software tool is used to crack passwords through a wordlist. Also, it can run in different operating systems such as Unix and Windows. Thus, by giving a limited key length and enough time, a brute force attack can be launched successfully on a ciphertext or plain text [11].

9.3.2 Implementation Attack

An implementation attack exploits the vulnerability of an application, software, or algorithms. Most of the encryption algorithms rely on outside resources of the algorithms. The implementation attack can breach some of the encrypted data due to weaknesses in the algorithms. There are three types of implementation attacks [1].

9.3.3 Side-Channel Attack

The attack is based on gathering vulnerabilities, a database of a specific target such as a company system, or a cryptography algorithm. Also, the side-channel attack can monitor all power's usage during encryption and decryption operations. This can lead to gain access to encrypted data such as algorithm keys. The most common cryptography algorithm that has been affected by the side-channel attack is AES [1].

9.3.4 Fault Analysis

The attack is based on forcing a target's system into an error situation to get wrong results of it. Usually, the attackers repeat the process until they gather a high amount of erroneous information. Finally, the attackers compare the results from the attack with previously known information of the target system; then, they analyze the information to gain sensitive information about the systems, which includes cryptography's algorithm data such as the secret key [12].

9.3.5 Probing Attacks

This attack is based on scanning a system, networks, and algorithms to discover vulnerabilities, which can be used to gain more privileges and steal sensitive information of a special target such as banking applications [13].

9.3.6 Birthday Attack

This is one of the most effective attacks, and it is used against the cryptographic hash. Birthday attack can lead to many potential effects such as disclosing sensitive information of a communication between two parties or more. For example, when students at a university asked about their date of birth, the answer of each student will be one of the 365 dates. Thus, if a student birthday is 3/8/199x, then we need 25 students to be asked. On the other hand, if a hash function has 64 bits of values, then the probability of hash values is 1.8×10^{19} [14].

In addition, if an attacker repeats the hash function for different values of inputs, a similar output will be after 5.1×10^9 of random values; so if an attacker finds two inputs having same hash value, then it will lead to break the hash function. Therefore, the birthday attack can lead to launch a collision, breaking hash functions and cryptography algorithms [14].

9.3.7 Chosen-Ciphertext Attack

A chosen-ciphertext attack is one of the model attacks that is used for cryptanalysis. The cryptanalysts gather different pieces of information to promote the attack successfully. Thus, adversaries use this type of attack to gather secret and important information about cryptography algorithms. The attack technique is based on selecting numbers of ciphertext to obtain its decryption versions by using a random or unknown key. The adversaries will launch the attack based on a set of ciphertext until the decrypted data appear. There are different number of cryptography schemes which have been attacked by the chosen-ciphertext attack such as RSA and SSL protocol [15].

9.3.8 Ciphertext-Only Attack (COA)

COA is one of the cryptographic attacks which is used for cryptanalysis. The attack is based on accessing only a set of ciphertext. The attackers are aiming to obtain the encrypted messages, information, or the secret keys of them to gain more data or other information in cryptography algorithms. Moreover, it may be hard to attack modern ciphers through COA. Thus, there are two methods that can help to breach the modern ciphers. The first method is exploiting the two-time pad. Cryptography experts say that never use the same keystream more than one time because the cipher will be able to attack by COA. The second method is frequency analysis, which is a common type of COA attack which focuses on analyzing the frequency of letters in a ciphertext. Each language or ciphertext will have its own letters that will make the frequency analysis to get successful results [1].

9.3.9 Power Analysis Attack (PAA)

This attack is based on gathering and abating information about an algorithm consumption in a way to determine the secret key or other sensitive data that can be used to decrypt the ciphertext. Moreover, this attack is similar to the timing attack, which is based on gathering facts of a process. For example, if an encryption process takes a long time, then the encryption algorithm uses a long secret key [16].

9.3.10 Known Plain Text Attack (KPA)

This attack is based on plain text. The attackers exploit the known plain text of some set of ciphertext. The aim is to use the known plain text to decrypt the set of ciphertext. This might be done by identifying the key or through other techniques. The most common example of this attack is linear cryptanalysis against block ciphers [17].

9.3.11 Chosen-Plain Text Attack (CPA)

This attack is based on a chosen ciphertext-plain text pair of an attacker choice. This means that the attacker has the plain text, which makes the attack process easier and is successful. Thus, the attacker aims to determine the encryption key of the cryptography algorithm. The CPA has been approved against different fields of cryptography such as hash functions, public-key cryptosystem, and RSA algorithm [18].

9.4 Cryptography Attacks Impacts

In general, any attack leads to leave an impact on the target due to many reasons. The cryptography algorithms have become vulnerable to numbers of cyber security attacks. Many aspects have been affected by the cyber security attacks against encryption algorithms, for example, information security breaches, financial losses, reputation, confidentially, integrity, and availability [1].

The information security is a significant aspect that should not be affected by any cyber-attacks. Each piece of information is playing an important role in the business field and different operations of sharing the information. Furthermore, losing a piece of information can lead to many issues, for example, if an organization is aiming to make a sale of a new technology product that does not exist yet. The new product has several futures and a specific price. Thus, if an attacker attacked this organization data center, they can steal this sensitive information and use it in malicious activities, for example, selling the secret information for other competitors in the market, which leads to many effects for the original organization. Therefore, the information should be secured to avoid similar impacts [19].

The financial loss is one of the most impacts that can affect a business situation directly. This means many organizations or governments will be affected economically due to many attacks. Thus, they cannot continue producing products or services for their clients and their operations will hold for a period of time due to this issue. This will affect employees' income, which can affect other aspects in their life such as health situation. In addition, if the financial issues continue, they may lead to bankruptcy, which is a huge impact on the business fields. This will decrease the operations of producing products or services, thus leading to loss of partnership with other companies [20].

Cryptography attacks such as brute force and man in the middle attack can leave a huge impact on an organization's reputation. The reputation plays a sensitive role in the market as it can attract new staff or investors. This can be done only if an organization has a strong name in the market such as never attacked by cyberattacks and has a professional staff in cyber security, operations, and recovery. On the other hand, if an organization has a poor name in the market, it will lose its revenue and name due to the cryptography breaches. The reputation of an organization can also be affected by other reasons such as economy issues or provides poor services, which make the customers to submit a complaint against the services. [21].

Confidentiality is a significant field that should be secured. It can be found in different areas such as business, health sector, immigration, and secret governments operations. Each of these sectors can be in a high risk and impacts if one of the cryptography attacks found on these sectors. For instance, the health sector contains a classified information of a client. The attacks can lead to reveal this information which should be protected and should have more privacy. Moreover, immigration sector also contains sensitive information such as passport numbers, visa numbers, and photographs of the passport holders. Any leaking on this type of data can breach the confidentiality of its users. Also, this sensitive information can be sold in black market, which can also be used in malicious activities under the passport's holders [22].

Furthermore, the following paragraphs explain more about the impacts of cryptography attacks:

9.4.1 Brute Force Attacks

The brute force attacks usually try to reveal hidden aspects of encrypted value or a plan password. Also, the attack can be used against numbers of encryption algorithms. Thus, the impacts which attack leave the following: Reveal sensitive information such as secret key value and plain text value which is usually a secret message between two parties and discolor of password values. These impacts actually will lead to other impacts of ozonizations or governments. For instance, government employees are sharing a piece of classified information regarding some business with other parties, but the lack of security in the used encryption algorithm such as DES leads to the loss of financial, loss of clients, and loss of reputation in the market [1]. Furthermore, the financial losses may be a reason that leads to other impacts. For example, an organization will lose its staff, because they are not paid enough and other competitors give them better offers. In situation, the losses will increase reputation, which is a sensitive issue [21].

9.4.2 Side-Channel Attacks

A side-channel attack can put cryptography algorithms users in high risks and can lead to some serious impacts. In addition, the most common algorithm that has been affected by the side-channel attack is the AES. The attackers use this kind of attacks which they are aiming to access the algorithm data to gain more information such as the secret key. Thus, once they got the secret key, other information became easier to reveal. This includes the transferred message or plain text data. Therefore, the side-channel attack is one of the strongest attacks, which leads to disclose the sensitive information [1].

9.4.3 Fault Analysis Attacks

A fault analysis attack affects the implementation of cryptosystem and its mathematical functions. For example, the elliptic curve cryptosystem has been proved that it is vulnerable to many cyber-attacks such as fault analysis attack. Thus, this type of attack exploits any fault that can happen in the implantation of the cryptosystems. For instance, faults can be exploited in many methods or locations, and almost the whole of the system can be attacked, such as the system parameters, dummy operations base point, intermediate results, dummy operations, and validation tests [13]. This leads to reveal and discover some secret information of the cryptosystem [12].

9.4.4 Probing Attacks

A probing attack is based on gathering sensitive information about a target. Usually, the gathered information will be vulnerabilities about the target by using specific tools and software. This kind of attack affects the target system by exploiting its weaknesses. For example, knowing the faults in a cryptography algorithm compromises it and steals all important data. The important data can be the secret key or the encrypted data. Thus, hackers exploit applications and system vulnerabilities to gain more access into it, and most organizations are vulnerable to probing attack if they are not updating their system or software [13].

9.4.5 Birthday Attacks

A birthday attack is one of the strong attacks in cyber security field. It is used to break cryptographic hash function such as MD5 or SHA hash function. This kind of hash function may contain some sensitive information such as a secret key of a cryptography algorithm or a password of a communication channel. Thus, the birthday attack can reveal sensitive information of a communication between two parties or more by attacking the cryptography process or hashes, which leads to the loss of the confidentiality and authentication of the communication channel. Moreover, the birthday attack can lead to launch hash collision to find the hash values. Also, breaks cryptography algorithms through [14].

9.4.6 Ciphertext-Only Attack (COA)

A COA, also known as known ciphertext attack, is a model attack for cryptanalysis, that is, the attackers can access only to a set of ciphertext. The attack can reveal the encryption or decryption key if the ciphertext has been decoded. This can be a risk for users who use cryptography as security features. Therefore, the loss of privacy, integrity, and authentication is possible if COA occurred [15].

9.4.7 Chosen-Ciphertext Attacks

A chosen-ciphertext attack leads to gather a secret or general information of cryptography algorithm. The attack is based on selecting numbers of ciphertexts to obtain its decryption versions by using a random or unknown key. From the decrypted versions, the attackers can recover or read some hidden aspects, such as the secret key which is used for encryption. One of the common examples that have been affected by CPA is the El Gamal cryptosystem. In fact, the El Gamal cryptosystem is secured against plain text attack, but it can be exploited by chosen-ciphertext attack. Also, the early versions of RSA use SSL protocol that was vulnerable to the chosen-ciphertext attack [15].

9.4.8 Known Plain Text Attack (KPA)

A KPA aims to affect an encryption scheme in a way to reveal or decode secret information. In addition, the attack is based on a set of plain texts and the encrypted version of them. Thus, the attackers try to use them to reveal sensitive information. The effects behind this attack are similar to those of COA and CPA, which reveal secret information and encrypted messages about an encryption algorithm [15].

9.5 Countermeasures

No one is immune in the cyber security field. Thus, the need for countermeasures is necessary to provide a high level of security to protect data of communication, online transactions, and e-commerce. All of these are using encryption processes to protect secret information against several breaches. The users of cryptography algorithms are required to select a secure method of encryption to ensure the security of their data. This can be through numerous techniques, tools, and policies while using an encryption scheme. Security and countermeasures will provide confidentiality, integrity, availability, and authentication of systems or sensitive data. In this chapter, the countermeasures are discussed in several methods [23].

Cryptography attacks prevention is not different from other cyber-attacks. The need in the cryptography field is to secure the encrypted data. This can be done by avoiding old encryption algorithms such as the DES, which is vulnerable to brute force attacks due to the small key length. Therefore, it is recommended to use secure encryption algorithms with a long key. For example, AES has different keys of 128, 192, and 256 bits. Also, it has a number of different rounds of encryption and decryption, so the key of 128 bits will be in 10 rounds, the key of 192 bits will be in 12 rounds, and the key of 256 bits will be in 14 rounds. Thus, the brute force attack cannot affect AES, due to the long keys used in the algorithm. Therefore, the usage of a strong encryption algorithm such as AES will ensure that the encrypted data cannot be breached [1].

The digital signatures play a significant role in securing identification and data. Using digital signature such as RSA while transferring data between the two parties, it will ensure the authentication of the two parties. In addition, the RSA algorithms use 1,024-bit and 2,048-bit key long, which make it more secure. In terms of cryptography concept, encryption of the data is not enough to make sure it is secured. Therefore, adding the digital signatures will add a new challenge to attackers, which increases their attack process [24].

In terms of securing encrypted data, it is recommended to implement information technology security policy such as ISO27001. This kind of policy will ensure several areas of security and increase awareness of the users such as how to choose system password [25]. Moreover, the awareness programs help to increase the cyber security awareness of users in organizations, banks, and business companies. This will aid to avoid many attacks such as phishing attacks. Phishing attacks can steal sensitive data from a user's device; this may include information of an encryption algorithm. Also, awareness programs aid to keep the organizations and others using cryptography futures in their environment in the latest update of security futures in cyber field such as encryption schemes. Thus, they will be aware of how to implement an encryption scheme securely [26].

9.6 Conclusion

Cryptography concept has become an essential thing in business field and government operations. This is to secure their data against breaches. Moreover, symmetric cryptography and asymmetric cryptography are used for different purposes in the cryptography field. One of the essential purposes is to ensure the confidentiality, integrity, and availability of their data. The symmetric cryptography has only one key to encrypt and decrypt a plain text. On the other hand, asymmetric has two keys used in encryption and decryption processes. Even though symmetric cryptography and asymmetric cryptography provide good security futures such as secure sensitive data, they can be attacked by several cyber-attacks. For example, a brute force attack takes advantage of the key length to launch the attack. Also, the birthday attack aims to break a hash function to steal its data. Moreover, there are other attacks that can be used to break symmetric cryptography and asymmetric cryptography, such as dictionary attack and COA, chosen-plain text attack, PAA, and side-channel attack. These attacks can break different types of encryption algorithms, and lead to reveal sensitive information of the algorithms such as private keys. Thus, this will lead to disclose sensitive information from the encrypted information. Therefore, the need of countermeasures is an essential thing to secure the users of cryptography algorithms. This chapter discussed the countermeasures by dividing it into prevention and deletion. The prevention section was referring to use unbreakable encryption algorithms such as AES scheme.

References

1. Gupta, A. & Walia, N. K. Cryptography algorithms: A review. *International Journal of Engineering Development and Research, 2*(2), 1667–1672, ISSN: 2321-9939.
2. Peter, S. & Dawn, T. (2019). Symmetric Key Encryption - Why, Where and How It's Used in Banking [Blog]. Retrieved from www.cryptomathic.com/news-events/blog/symmetric-key-encryption-why-where-and-how-its-used-in-banking.
3. Al-Vahed, A., & Sahhavi, H. (2011). An overview of modern cryptography. *World Applied Programming, 1*(1), 55–61.
4. Shree, D., & Ahlawat, S. (2017). A review on cryptography, attacks and cyber security. *International Journal of Advanced Research in Computer Science, 8*(5), 1–4.
5. Paar, C., & Pelzl, J. (2010). *Understanding Cryptography: A Textbook for Students and Practitioners*. Heidelberg: Springer. Retrieved from https://ecu.on.worldcat.org/search?databaseList=&queryString=Understanding+cryptography%3A#/oclc/527339793.
6. Paliwal, S., & Gupta, R. (2012). Denial-of-service, probing & remote to user (R2L) attack detection using genetic algorithm. *International Journal of Computer Applications, 60*(19), 57–62.
7. Köpf, B., & Smith, G. (2010, July). Vulnerability bounds and leakage resilience of blinded cryptography under timing attacks. In *2010 23rd IEEE Computer Security Foundations Symposium* (pp. 44–56). IEEE, Edinburgh, UK.
8. Conrad, E. (n.d.). Types of Cryptographic Attacks. Retrieved from www.academia.edu/4739047/Types_of_Cryptographic_Attacks.
9. Adams, C. (2011). *Dictionary Attack* (pp. 1–76). School Of Information Technology and Engineering (SITE), University Of Ottawa. doi:10.1007/978-1-4419-5906-5_74.
10. Kalenderi, M., Pnevmatikatos, D., Papaefstathiou, I., & Manifavas, C. (2012, August). Breaking the GSM A5/1 cryptography algorithm with rainbow tables and high-end FPGAS. In *22nd International conference on field programmable logic and applications (FPL)* (pp. 747–753). IEEE, Oslo, Norway.

11. Weir, M., Aggarwal, S., De Medeiros, B., & Glodek, B. (2009, May). Password cracking using probabilistic context-free grammars. In *2009 30th IEEE Symposium on Security and Privacy* (pp. 391–405). IEEE, Berkeley, CA.

12. Joye, M., & Tunstall, M. (Eds.). (2012). *Fault Analysis in Cryptography* (Vol. 147). Heidelberg: Springer.

13. Saraswat, S., Tanisha S., & Neetu F. (2017). Fault analysis in cryptography. *International Journal of Latest Trends in Engineering And Technology.* doi:10.21172/1.91.29.

14. Gupta, G. (2015). *What is Birthday Attack??* (pp. 1–14). The Maharaja Sayajirao University of Baroda. Retrieved from www.researchgate.net/publication/271704029_What_is_Birthday_attack.

15. Sadkha, S. (2012). Methods of Cryptanalysis. Retrieved from www.uobabylon.edu.iq/eprints/paper_5_7264_649.pdf.

16. Ors, S. B., Gurkaynak, F., Oswald, E., & Preneel, B. (2004, April). Power-analysis attack on an ASIC AES implementation. In *International Conference on Information Technology: Coding and Computing, 2004. Proceedings. ITCC 2004* (Vol. 2, pp. 546–552). IEEE, Washington, DC.

17. Van Oorschot, P. C., & Wiener, M. J. (1990, May). A known-plaintext attack on two-key triple encryption. In *Workshop on the Theory and Application of of Cryptographic Techniques* (pp. 318–325). Springer, Berlin, Heidelberg.

18. Kiltz, E., O'Neill, A., & Smith, A. (2010, August). Instantiability of RSA-OAEP under chosen-plaintext attack. In *Annual Cryptology Conference* (pp. 295–313). Springer, Berlin, Heidelberg.

19. Zhang, D. (2018, October). Big data security and privacy protection. In *8th International Conference on Management and Computer Science (ICMCS 2018).* Atlantis Press, Shenyang, China.

20. Bouveret, A. (2018). Cyber Risk for the Financial Sector: A Framework for Quantitative Assessment. International Monetary Fund.

21. Mitic, P. (2018). Reputation risk: Measured. *International Journal of Safety and Security Engineering, 8*(1), 171–180.

22. Levenstein, M. C., Tyler, A. R., & Davidson Bleckman, J. (2018). The Researcher Passport: Improving Data Access and Confidentiality Protection.

23. Ghosh, S., Mukhopadhyay, D., & Chowdhury, D. R. (2011). Fault attack, countermeasures on pairing based cryptography. *IJ Network Security, 12*(1), 21–28

24. Merkle, R. C. (1989, August). A certified digital signature. In *Conference on the Theory and Application of Cryptology* (pp. 218–238). Springer, New York.

25. Disterer, G. (2013). ISO/IEC 27000, 27001 and 27002 for Information Security Management.

26. Abawajy, J. (2014). User preference of cyber security awareness delivery methods. *Behaviour & Information Technology, 33*(3), 237–248.

Chapter 10

Cyber Security for Network of Things (NoTs) in Military Systems: Challenges and Countermeasures

Muhammad Imran Malik, Ian Noel McAteer, Peter Hannay, and Ahmed Ibrahim
Edith Cowan University

Zubair Baig
Deakin University

Guanglou Zheng
Edith Cowan University

Contents

10.1 Introduction

The proliferation of automated solutions for the commercial and the private domain at an unprecedented rate, coupled with ever-increasing geopolitical threats being faced by nation states, has led to the adoption of technological solutions by the military to handle complex battlefield situations. Pradhan et al. [1] argue that today's highly dynamic and unpredictable military operations require constant technology renewal and deployment, which has compelled the necessity of adopting the Network of Things (NoT) by militaries across the world. Voas [2] defines NoT as a discipline, of which Internet of Things (IoT) is an instance, and argues that the difference between the two terms is quite subtle. Further argument by Voas [2] elucidates IoT as devices connected to the Internet and NoT as devices not connected to the Internet. As militaries tend to have their cloud/ networks parallel to the general Internet, the use of NoT stands more appropriate than IoTs. Hence, this paper has used the term "NoT," which is more relevant in a military context.

Deployment of NoTs is transforming the way modern militaries communicate, collaborate, and coordinate through their command and control systems installed on board their setups/equipment. NoTs are also being extensively employed in Industrial Control Systems (ICS) such as Supervisory Control And Data Acquisition (SCADA) networks being used in the military environment to monitor and control critical infrastructure comprising automated processes [3]. Though still in infancy, modern militaries have started the adoption of NoTs to manage some assets and coordinate complex and distributed systems, as it [NoTs] augments the performance of equipment, improves efficiency, and ensures the safety of men and material. Pradhan et al. [4] argue that increased dependence on Network-Centric Warfare (NCW) capabilities that integrates physical, information, and cognitive domains has also led to the adoption of NoT-related technologies in key areas of military defense.

The concept of joint operations put forth by developed countries envisions that by 2020, digital collaboration technologies, including NoTs, will enable distributed commanders and staffs to collaborate as if they are co-located [5]. By embedding

NoT-enabled devices and applications in military equipment, the commanders will be able to use the data gleaned from the devices flawlessly in the form of Common Operational Picture (COP) [4] to handle complex war scenarios effectively and proactively. However, this comes with various security, safety, and interoperability challenges. On the combat side in a military setting, Zheng and Carter [6] state that the rapid growth of NoT is driven by C4I or C4ISR systems. These systems use millions of sensors [6] deployed on a range of platforms to provide situational awareness to decision-makers and warfighters deployed in the headquarters enabling them to take quick decisions. Not only this, but NoTs are also paving its way in fire-control systems [7], an integral part of any military defense, through the end-to-end deployment of networked sensors with pinpoint precision. NoT is also making its way in predictive maintenance of equipment through the use of real-time data from various kinds of equipment to determine when repairs should take place before breakage occurs [6]. On the non-combat side, NoTs also improve the efficiency and effectiveness of back-end processes such as tracking shipments and managing inventories between central logistics hub, training, and simulation exercises, etc.

> Every age has its own kind of war, its own limiting conditions, and its own peculiar preconceptions [8].

The picture portrayed above clearly indicates that adoption of NoTs in the short term will witness a significant influx in the military environment. NoTs, therefore in years to come, will proliferate in military equipment such as tanks, ships, submarines, and fighter jets, as well as various industrial systems and other non-combat military applications. This ecosystem will create its cyberspace with associated challenges/risks—security being the paramount challenge to broader IoT adoption across combat and non-combat systems. The situation becomes grave when commercial off-the-shelf (COTS) IoT devices and applications are used as NoT devices "as is" on the pretext of enabling faster technology acquisition and deployment in the military [1]. Arias et al. [9] argue that security of such NoT devices and applications has predominantly taken a backseat by their vendors in a rush to be first to the market with a new product. Consequently, security measures required for foolproof and attack-resistant devices remain largely unaddressed. The situation, therefore, demands that greater dependence on the use of NoT technology by military be dealt critically by analyzing the underlying architecture used in such devices and applications. Such scrutiny may include an analysis of protocols and encryption mechanisms as well as data collection, distribution, feedback, and analytical frameworks from a security perspective to better safeguard national interest of any sovereign state.

The objective of this research is to analyze the effectiveness and efficacy of deployment of NoT devices and applications in military systems that are now considered inevitable for increased situational awareness by military commanders. Furthermore, the research also aims to examine various security challenges associated with NoTs mainly due to the lack of awareness and poor design approach by NoT manufacturers. Following is our list of contributions through this research work:

1. Elaboration upon the effective use of NoTs by militaries to manage complex and distributed operations both in combat and in non-combat setting to gain information superiority,
2. Description of various security challenges associated with the use of NoTs in a military environment, and
3. Proposal of effective mitigation strategies for the identified cyber security challenges.

10.2 Background Information

The following timeline enumerates the exponential growth in NoT usage over recent years and anticipated growth into the near future:

1. 1998—IPv6 protocol, which allows an address space of 2^{128} unique IP addresses, formalized by the Internet Engineering Task Force (ITEF) to prepare for the prediction that the IPv4 address space will become exhausted [10].
2. 1999—The term "IoT" originally coined [11].
3. The early 2000s—The first commercial domestic appliance IoT device, the Internet refrigerator, began being shipped by some manufacturers [12].
4. 2008—The number of devices connected to the Internet is considered to exceed the number of people on earth [13].
5. 2017—Gartner estimates 8.4 billion NoT devices in existence [14], whereas IHS (the Information Handling Services) Markit estimates to the figure is 20 billion [15].
6. 2020—Gartner estimates 20 billion NoT devices [14], whereas Juniper Research estimates that it could be as much as 38 billion [16].

Not only has the number of NoT devices expanded, so has the range of applications into which they are being implemented. However, the "first-to-market" race undertaken by NoT developers has resulted in issues related to device security given scant regard [17].

The infrastructure that makes up everything encompassed within an NoT system goes far beyond the NoT devices themselves. These devices are positioned at one end of a chain of two-way communications that extend to a centralized

Figure 10.1   NoT architecture.

back-end data system. TechBeacon [18] defined this communication chain as being a four-stage process:

1. NoT sensors and actuators devices on the perimeter acquiring data to be monitored.
2. Systems performing sensor data aggregation and analogue-to-digital conversion.
3. Edge IT systems which perform pre-processing of data and transmission.
4. The back-end data systems where data analysis, management, and storage are performed.

The two-way communication comes from acquired data being transmitted in one direction to the back-end data system, while command and control communications from the back-end data system are transmitted through the same chain back to the NoT devices (Figure 10.1).

10.3 Discussion

10.3.1 Network-Centric Warfare (NCW)

Publications in the late 1990s/early 2000s for the U.S. Department of Defense identified three integrated domains which composed the core concepts of NCW [19–21], namely:

1. Physical domain—The place where data is generated from sensors and human observers.
2. Information domain—The transmission and storage of the generated data.
3. Cognitive domain—The processing and analysis of the received data.

Zheng and Carter [6] argue that more than a decade later, these three domains still apply in that military NoT usage to date has, for the most part, targeted Command, Control, Communications, Computers, Intelligence, Surveillance, and Reconnaissance (C4ISR) and fire-control systems. However, military NoT usage has not always predominantly focused only on passive information-gathering and forwarding of this information for analysis at command centers. For several years, certain items of military equipment that can be considered to be NoTs have either passive or offensive capabilities.

10.3.2 Present-Day Passive Military NoT Devices

> If you know the enemy and know yourself, you need not fear the result of a hundred battles. If you know yourself but not the enemy, for every victory gained you will also suffer a defeat. If you know neither the enemy nor yourself, you will succumb in every battle [22].

Throughout history, commanders who are in possession of the greater amount of accurate information before a conflict commences, invariably win. For the human race, never has there been such quantities of readily available information than at present following the development of the Internet. The evolution of the proliferation and diversity of network-centric information-gathering sensors gives the better-equipped military force the advantage of information-dominance during any conflict.

For several years, unmanned aerial vehicles (UAVs or drones) have been high-altitude eyes-in-the-sky with inbuilt high-definition optics. They provide detailed live video feeds or still photographs over an area of operations, are silent and invisible to those on the ground below, remain largely undetectable to surface-to-air missiles, and avoid putting a pilot's life in danger. They are considerably cheaper than an equivalent manned reconnaissance aircraft and can be piloted remotely from a command and control center anywhere in the world. These advantages have been recognized since the early days of UAV assimilation into the military field of operations [23].

Similar to several law-enforcement agencies around the world, body-mounted or helmet-mounted cameras are becoming part of a soldier's combat uniform. Such live video access to frontline action can be viewed remotely when the footage is streamed wirelessly via, perhaps, an overhead drone, as proposed by Kim et al. [24].

Augmented reality (AR) superimposes computer-generated elements into real-world sensory inputs of sight, sound, and data. Head-up displays have assisted air force pilots for several years by overlaying their field of vision with flight and fire-control information. Such technology is now available for ground troops, whose AR goggles can provide location, mapping, and friend/foe identification information [25].

The adoption of RFID (Radio Frequency IDentification) technology within the military has been slow, perhaps due to a perceived poor return on investment [26]. However, these misconceptions are gradually being overcome, and the use of the technology for asset tracking across the armed forces is gaining momentum. Being able to track the location and inventory of equipment, spare parts, ammunition, uniforms, fuel, water and food supplies, etc. via RFID scanners and online databases is considerably more efficient than the previous manual counting and paper-trail methodology.

Another form of asset tracking at a unit level is at the forefront of existing NoT sensor data-gathering technology. ALIS (Autonomic Logistics Information

System) is currently unique to the Lockheed Martin F-35 Lightning II joint strike fighter aircraft [27]. Embedded sensors throughout the F-35 monitor the performance of components and compare these to pre-set parameters. Maintenance needs are then predicted by analytical software, which in turn ensures that the required spare parts are delivered to maintenance staff in time for when they are needed [26]. While the system has reportedly had some technical glitches since its implementation, it is an example of how NoT technology can augment the efficiency of logistics processes.

10.3.3 Present-Day Offensive Military NoT Devices

All warfare is based on deception. Hence, when we are able to attack, we must seem unable; when using our forces, we must appear inactive; when we are near, we must make the enemy believe we are far away; when far away, we must make him believe we are near [22].

The development of NoT devices in the military has made possible both over-the-horizon and invisible-from-above deception capabilities while achieving sub-meter strike accuracy.

Tomahawk cruise missiles are NoT devices, networked to GNSS (Global Navigation Satellite System) satellites to fly a programmed route according to the three-dimensional (3D) topography map in its memory while avoiding enemy radar. They can transmit video footage from a camera in its nose cone right up to the moment of impact on its assigned target via a communications satellite or terrestrial means. They can be deployed from truck-mounted launchers, naval ships, submarines, or aircraft. No one who watched the TV news coverage of the first Gulf War in the early nineties will forget the images of cruise missiles passing over the streets of Baghdad as they approached their designated target or the onboard footage of the strike itself. The latest generation of cruise missiles is now capable of, for example, taking and relaying reconnaissance photographs from one location before being reprogrammed mid-flight to fly to an alternative target [28].

UAVs have developed from the passive reconnaissance role described above to becoming armed (strike-enabled UAVs) and capable of launching missiles. In the early 2000s, the MQ-1 Predator drone's payload capacity restricted armaments to two AGM-114 laser-guided Hellfire missiles [29]. Currently, larger drones such as the MQ-9 Reaper ("Predator B") and Avenger ("Predator C") can carry maximum payloads of 1,747 kg [30] and 2,948 kg [31], respectively. They are designed to carry armaments in various configurations, including up to four of the latest generation of Hellfire missiles. Ironically, this development has raised considerable ethical debate on whether killing from a risk-free standpoint constitutes war at all [32].

10.4 Challenges and Countermeasures for Future Military Warfare

Military organizations around the world are becoming more and more aware of the benefits that NoT devices can bring to a variety of sectors within the military field of operations. These sectors include command and control center operations, front-line and asset surveillance, support logistics, and field medical care [33]. The same technology that will transform our conurbations into "smart cities" will transform areas of conflict into "smart battlefields" (Figure 10.2).

To achieve assimilation between potentially thousands of NoT sensors, the transmission and storage of vast amounts of data will be required, even to the exabyte level [34]. Such developments into the future will continue and become viable for increasing network sizes, just as NoT communication capacity has developed in recent years within the domestic, education, and industrial sectors.

In their recent report, Allen and Chan [35] propose that artificial intelligence (AI) is an emerging technology that will improve cyber security, the accuracy of firing solutions for weapons systems, and so on in the military sector. The development of autonomous vehicles is already underway to supplement the hazardous tasks taken by bomb-disposal squads, thereby helping to reduce risk to human life [36].

In 2016, NATO officially recognized cyberspace as being the "fifth domain" of warfare after land, air, sea, and space [37]. Consequently, any cyber-attack on a

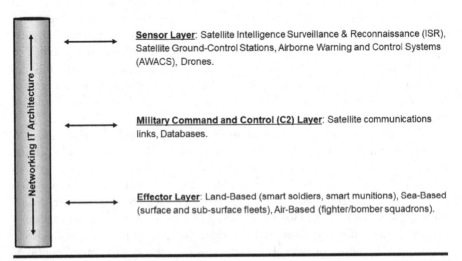

Figure 10.2 Future military network-based operations.

nation state would be considered as an act of war. In such a scenario, the frontline moves from the battlefield to a computer console, and from a soldier to a black-hat hacker intent on penetrating the IT infrastructure of an adversary state. The diversity of the damage that well-targeted cyber-attacks could achieve is as eclectic as the minds of evil men are capable:

1. Compromising dams to release reservoirs and cause widespread flooding,
2. Compromising power stations to cause widespread blackouts,
3. Compromising flight management systems to bring down, in particular, government or military aircraft,
4. Launch a nation's nuclear weapons against itself,
5. Compromise traffic-control systems to cause gridlock throughout conurbation centers.

Within the digital transformation through NoTs in military systems comes the significant requirement to ensure that the NoT devices and their lines of communication are encrypted and secured against any potential cyber-attack.

10.5 Cyber Security Challenges

Increased dependence on network-centric operations integrating physical, information, and cognitive domains fits exceptionally well within the definition of distributed technologies. That involves hardware devices, software, as well as the networks. Therefore, network-centric systems are prone to a vast array of cyber-attacks mainly because the security of the NoT devices and applications is not keeping up with the pace of innovation by the manufacturers [38]. Knowingly that public and private organizations have seen a significant influx of network-centric capabilities both at the operational and at the support side, the proliferation of increased dependence on technology enablers has created a cyberspace, which has significant challenges/risks associated with it. Among these risks, security is considered as the paramount challenge across combat and non-combat systems. The rapid evolution of cyber threat landscape has also seen such threats moving from traditional attacks to Advanced Persistent Threats (APT) attacks through which integrated networks are becoming more and more vulnerable. Like in any other computing system, NoT security should be compliant with essential components of information assurance, i.e., confidentiality, integrity, availability, authentication, and non-repudiation. While fulfilling the requirements to comply with the pillars of information assurance, a few of the common challenges/issues that are applicable in a military setting while using NoTs are succinctly discussed in Table 10.1.

Table 10.1 Cyber Security Challenges with NoTs in Military Systems

Folio	Vulnerability	Challenges
1	Reliance on vendor design (physical security attacks)	NoT products employed are often installed without understanding its hardware architecture/design. While the device may fulfill the purpose, not considering target system's requirement may give an attacker the exposure of interfaces that can be compromised leading to leaking of sensitive information or installation of malware to control the device [9]
2	Poor encryption mechanism	Inadequate encryption strength allows an adversary to exploit this weakness and decrypt the data at rest or while in motion. Furthermore, weak cryptographic implementation also has the potential of misusing faulty SSL/TLS implementation by spoofing SSL/TLS servers via arbitrary certificates or by allowing firmware installation remotely by the attacker by generating a similar attack on the distribution server [9]
3	Unpatched software/firmware (operating system and application integrity attacks)	Whether an NoT device uses an open or proprietary software, not regularly updating the software allows an attacker to exploit and control the device through different attack vectors [9]
4	Electronic warfare (e.g., denial of service attack)	Since NoT devices communicate wirelessly especially when deployed as sensors, an adversary can use techniques such as RF (radio frequency) jamming to block the signals making devices unable to share data with base stations [6]. Wireless signals also have the potential to expose position/location of a soldier or a vehicle [6]. Furthermore, if the antenna power used is too weak, the intended destination for the transmitted signals will have no, or at best intermittent, reception. If the antenna power used is too high, then the transmitted signal will be broadcast far beyond its intended destination, thereby increasing the probability of signal interception by enemy forces

(Continued)

Table 10.1 (*Continued*) Cyber Security Challenges with NoTs in Military Systems

Folio	Vulnerability	Challenges
5	Insider threat/ compromise	Brodie [39] claims that the weakest link in any organization's security is its people. Such weaknesses can be derived from simple ignorance concerning information security; social engineering via pretexting, phishing, or baiting; or they can be threatened, bribed, or blackmailed. Zheng and Carter [6] argue that a single mistake from an employee gives an attacker access to the device(s) using relatively naive means
6	Interoperability	A large number of systems designed by different manufacturers are used in the military for offensive and defensive capabilities. While it is based on vendor's expertise, not giving enough consideration of having common set of standards and protocols restricts the NoT devices to talk to each other [6]. This issue has the potential of not to accrue the benefits of NoT and gives adversary enough time to employ a strategy, especially during wartime
7	Wired/wireless scanning and mapping attacks	A skilled adversary will always attempt to gain access to one's network, via exploiting any identified vulnerability. Scanning and locating open ports represent one of the more direct methods to achieve this goal. Attention to detail and vigilance is required to ensure a hardening of one's network to deter such attacks [40]
8	Protocol compromise	Cross-protocol attacks are a new attack vector that uses the strength in one protocol against weakness in another protocol. The commonly used SSL encryption system and its successor TLS are both susceptible to this kind of attack [41]. For example, DROWN (Decrypting RSA with Obsolete and Weakened eNcryption) is a recent attack that exploits HTTPS and other services that use SSL/TLS [41,42]
9	Eavesdropping .	Eavesdropping is a passive man-in-the-middle attack whereby the threat entity intercepts and deciphers transmissions to learn of their content, but does not take any active attack measures on the network compromised. In the military context, this may be to learn of an adversary's tactical plans [43]

(Continued)

Table 10.1 (*Continued*) Cyber Security Challenges with NoTs in Military Systems

Folio	Vulnerability	Challenges
10	Spoofing and masquerading	These are both active man-in-the-middle attacks whereby the threat entity can substitute information to give the appearance that he is on his adversary's side. Such aims could be achieved through the insertion of false yet plausible information or changing his MAC address to one that is known to be on his adversary's network [44]
11	Real-time monitoring of network traffic	Clausewitz [8] stated: "all action takes place, so to speak, in a kind of twilight, which, like fog or moonlight, often tends to make things seem grotesque and larger than they really are." It is from this that the term "fog of war" was derived. Prahbhu et al. [45] discuss field noise, field variation, and background signals as all being issues within a combat system. Such "fog" serves to obfuscate the ability to achieve real-time network traffic monitoring

10.6 Proposed Countermeasures

For militaries, network-centric systems are aimed at increasing the efficiency through assured command and control networks that are uncontested and therefore follow multipath, multichannel network, following the Sun Tzu's concept of deception. However, inadequate security of these systems allows an adversary with intelligence to launch various cyber-attacks—a subset of which is discussed earlier. Therefore, a military NoT network should comply with the five basic principles of security, namely, layering, limiting, diversity, obscurity, and simplicity [46]. Such an approach is called "Defense-in-Depth" (DiD), which is the concept of protecting computer network with a series of defensive mechanisms such that if one mechanism fails, another will already be in place to thwart an attack [47,48]. DiD can be better achieved by reducing the likelihood or the severity of an attack; security controls need to be well placed to restrict an adversary's ability to exploit a vulnerability. Figure 10.3 illustrates the need of having controls for DiD approach and their relationship with attacks and vulnerabilities.

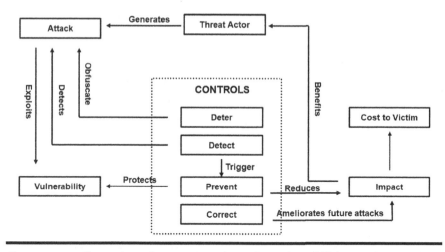

Figure 10.3   Illustration of ecosystem of attacks, vulnerabilities, and controls.

To address security challenges being faced when NoTs are employed, the following principles are considered essential for improving the security posture of military systems (Table 10.2):

Table 10.2   Strategic Principles to Address NoT Security Challenges [38]

Folio	Principles	Objectives
1	Incorporate security at the design phase	Reduces the potential disruptions and helps in avoiding the endeavor of adding the security features to NoT devices once they are deployed
2	Develop a strategy for advance security updates and vulnerability management	Addresses the flaws highlighted after deploying an NoT device
3	Build on recognized security practices by using traditional IT and network security policies	Such approach would help in finding the irregularities, responding to potential incidents, etc.

(Continued)

Table 10.2 (*Continued*)   Strategic Principles to Address NoT Security Challenges [38]

Folio	Principles	Objectives
4	Prioritize security measures according to potential impact	A risk model needs to be developed that should address what type of information is being handled by each NoT device and its potential impact in case a security breach occurs. This includes but not limited to defining intended use of the device, performing red-teaming exercises, identification and authentication of devices within the network, etc.
5	Promote transparency across NoT	Vendors and military users need to deliberate the inherent architecture of the devices in use and continue this process even after the NoT devices are deployed and in use. This will increase awareness of both the stakeholders and help in applying security measures more efficiently and help in an efficient incident response mechanism
6	Connect carefully and deliberately	The risks associated with the NoT devices can cause serious damage as discussed throughout this paper. It is advisable to understand the need of NoT device being deployed and which other devices it can connect. A mechanism as when an NoT device should share data with other devices within the network also needs to be devised to reduce the security threats

Further to the above recommendations, the 20 Critical Security Controls (CSC) promulgated by The Center for Internet Security (CIS) for specific and actionable ways to protect an organization from the most pervasive and protective attacks are also considered effective for use in NoT-enabled military systems. These controls are aimed at effective cyber defense and are regularly updated based on the ever-increasing cyber threat landscape. Wade [49] argues that CSC controls to form a perfect baseline for an organization aiming to improve their information security posture as these controls have been developed from the contributions of some government and private sponsors, forensic experts, and penetration testers. The CIS also mapped these security controls with renowned security frameworks such as NIST Cybersecurity Framework, ISO 27002:2013 and 27002:2005. Also used were the security critical controls/mitigations strategies promulgated by National

Security Agency (NSA), Government Communications Headquarters (GCHQ), UK Cyber Essentials, Australian Signals Directorate (ASD) Top 35 mitigation strategies, etc. [50].

The shortage of a skilled workforce in the cyber-domain is another factor that contributes towards undermining the otherwise revolutionary capabilities of NoTs in military systems [51]. This fact can also be applied to the security of cyber-smart militaries. While the long-term solution to the issue is the effective implementation of cyber careers in militaries, the crux and vital ingredients are to engage the private sector and academia in improving the security posture. Two such examples towards this approach have already been initiated by the Australian government by partnering with the industry and academia to build research and workforce capabilities in cyber security. The establishment of Academic Centres of Cyber Security Excellence (ACCSE) [52] and industry-led Cooperative Research Centres (CRC) for cyber security [53] are considered efficient and swift solutions towards addressing the continuously evolving cyber security challenges.

10.7 Conclusion

Cyber threats are events that have the potential to adversely impact the organization's operations via unauthorized access, destruction, disclosure, modification of information, and (distributed) denial of service attacks. These threats can be physical or logical, to the internal and external components, directed or non-directed in nature, and have the potential for direct or indirect adverse effects or consequences. Greater dependence on operations like surveillance, reconnaissance, and use of drones for effective military operation by keeping an eye on the adversary's movements has paved the way for the use of NoTs by the militaries. However, these technology enablers have transformed the cyber threat landscape as modular and multifaceted in which the adversaries steadily and successfully develop tools that defy the security controls installed in network-centric systems. As such, a network-centric environment is vulnerable to various cyber-attacks, and all security mechanisms (tools, practices) in place need to be reviewed constantly and dynamically. Adoption of proactive approach against the ever-growing cyber threat landscape is considered as the only possible way for a network-centric entity to protect or defend the use of its cyberspace and infrastructure from cyber-attacks.

This research has attempted to discuss the effectiveness of having a network-centric capability by the modern militaries considered essential in this technological era to gain information superiority. Various security challenges associated with NoTs have also been elaborated at length in the paper. The countermeasures presented, if followed diligently, can reduce these challenges, and help improve the security posture of the military, which in turn will better protect the matters of national interest.

10.8 Future Work

The research undertaken for this paper presented a high-level view of how NoTs can be exploited by modern militaries to accrue maximum benefits that would assist commanders in achieving the assigned mission objectives. As part of the future work, the authors intend to conduct an in-depth review of various risks associated with non-deployment of security controls in NoTs and how game-theoretic approach can be employed for the commander-adversary attack model, wherein, if the commander does indeed stay ahead in the game, wins!

Abbreviations

The following abbreviations are used in this manuscript:

ACCSE	Academic Centres of Cyber Security Excellence
ALIS	Autonomic Logistics Information System
APT	Advanced Persistent Threat
AR	Augmented Reality
ASD	Australian Signals Directorate
C4I	Command Control Communications Computers and Intelligence
C4ISR	Command Control Communications Computers Intelligence Surveillance and Reconnaissance
CIS	Center for Internet Security
COP	Common Operational Picture
COTS	Commercial Off-the-Shelf
CRC	Cooperative Research Centre
CSC	Critical Security Controls
DiD	Defense-in-Depth
DROWN	Decrypting RSA with Obsolete and Weakened eNcryption
GCHQ	Government Communications Headquarters
GNSS	Global Navigation Satellite System
HTTPS	Hyper Text Transfer Protocol Secure
ICS	Industrial Control Systems
IoT	Internet of Things
MAC	Media Access Control
NCW	Network-Centric Warfare
NSA	National Security Agency
NoT	Network of Things
RFID	Radio Frequency IDentification
SCADA	Supervisory Control And Data Acquisition
SSL	Secure Sockets Layer
TLS	Transport Layer Security
UAV	Unmanned Aerial Vehicle

References

1. M. Pradhan, F. Gökgöz, N. Bau, and D. Ota, "Approach towards application of commercial off-the-shelf Internet of Things devices in the military domain," in *Internet of Things (WF-IoT), 2016 IEEE 3rd World Forum, Reston, VA, USA*, 2016: IEEE, pp. 245–250.
2. J. Voas. (2016). *Networks of 'Things'* [Online]. Available: http://nvlpubs.nist.gov/nistpubs/SpecialPublications/NIST.SP.800-183.pdf.
3. J. Romero-Mariona, R. Hallman, M. Kline, J. San, M. Major, and L. Kerr, *Security in the Industrial Internet of Things: The C-SEC Approach*, 2016.
4. M. Pradhan, A. Tiderko, and D. Ota, "Approach towards achieving an interoperable C4ISR infrastructure," in *Military Technologies (ICMT), 2017 International Conference, Brno, Czech Republic*, 2017: IEEE, pp. 375–382.
5. CCJO, "Capstone Concept for Joint Operations (CCJO): Joint force 2020," US Department of Defense, Joint Electronic Library, 2012, [Online]. Available: www.dtic.mil/doctrine/concepts/concepts.htm.
6. D. E. Zheng and W. A. Carter, *Leveraging the Internet of Things for a More Efficient and Effective Military*. Lanham, MD: Rowman & Littlefield, 2015.
7. T. SimplexGrinnell. (2015, September 17). *Leveraging Internet of Things technology to improve fire and life-safety protection [Video file]* [Online]. Available: www.youtube.com/watch?time_continue=14&v=NNZwfcN62z0.
8. C. v. Clausewitz, *On War*. Princeton, New Jersey: Princeton University Press, 1989.
9. O. Arias, J. Wurm, K. Hoang, and J. Yier, "Privacy and security in Internet of Things and wearable devices," *IEEE Transactions on Multi-Scale Computing Systems*, vol. 1, no. 2, pp. 99–109, November 06, 2015.
10. S. E. Deering. (1998). *Internet protocol, version 6 (IPv6) specification* [Online]. Available: https://tools.ietf.org/html/rfc2460.
11. K. Ashton. (2009). *That 'internet of things' thing* [Online]. Available: www.rfidjournal.com/articles/view?4986.
12. F. Osisanwo, S. Kuyoro, and O. Awodele, "Internet refrigerator – A typical internet of things (IoT)," *Presented at the 3rd International Conference on Advances in Engineering Sciences & Applied Mathematics (ICAESAM'2015)*, London (UK), 2015. [Online]. Available: http://iieng.org/images/proceedings_pdf/2602E0315051.pdf.
13. D. Evans. (2011). *The internet of things [Infographic]* [Online]. Available: http://blogs.cisco.com/diversity/the-internet-of-things-infographic.
14. R. v. d. Meulen. (2017). *Gartner says 8.4 billion connected "Things" will be in use in 2017, up 31 percent from 2016* [Online]. Available: www.gartner.com/newsroom/id/3598917.
15. P. Brown. (2017). *20 billion connected Internet of Things devices in 2017, IHS Markit says* [Online]. Available: http://electronics360.globalspec.com/article/8032/20-billion-connected-internet-of-things-devices-in-2017-ihs-markit-says.
16. S. Smith. (2017). *'Internet of Things' connected devices to almost triple to over 38 billion units by 2020* [Online]. Available: www.juniperresearch.com/press/press-releases/iot-connected-devices-to-triple-to-38-bn-by-2020.
17. J. Gubbi, R. Buyya, S. Marusic, and M. Palaniswami, "Internet of Things (IoT): A vision, architectural elements, and future directions," *Future Generation Computing Systems*, vol. 29, no. 7, pp. 1645–1660, 2013.

18. TechBeacon. (2017). *The 4 stages of an IoT architecture* [Online]. Available: https:// techbeacon.com/4-stages-iot-architecture.

19. D. S. Alberts, J. J. Garstka, and F. P. Stein, *Network Centric Warfare: Developing and Leveraging Information Superiority*. Washington, DC: CCRP, 1999.

20. D. S. Alberts, J. J. Garstka, R. E. Hayes, and D. A. Signori, *Understanding Information Age Warfare*, Washington, DC: CCRP Publication Series 2001.

21. D. S. Alberts and R. E. Hayes, *Power to the Edge: Command and Control in the Information Age*. Washington, DC: CCRP, 2003.

22. S. Tzu, *The art of war*, 500BC.

23. D. Glade. (2000). *Unmanned aerial vehicles: Implications for military operations* [Online]. Available: www.dtic.mil/get-tr-doc/pdf?AD=ADA425476.

24. S. J. Kim, G. J. Lim, and J. Cho, "Drone relay stations for supporting wireless communication in military operations," in *Advances in Human Factors in Robots and Unmanned Systems: Proceedings of the AHFE 2017 International Conference on Human Factors in Robots and Unmanned Systems, July 17–21, 2017, The Westin Bonaventure Hotel, Los Angeles, California, USA*, J. Chen, Ed. Cham: Springer International Publishing, 2017, pp. 123–130.

25. D. Ergürel. (2017). *Augmented reality meets military service* [Online]. Available: https://haptic.al/us-army-augmented-reality-2191b7e1856a.

26. J. Mariani, B. Williams, and B. Loubert. (2015). *Continuing the march: The past, present, and future of the IoT in the military.*

27. Lockheed Martin. (2017). *Autonomic logistics information system* [Online]. Available: www.lockheedmartin.com.au/us/products/ALIS.html.

28. C. Drubin, "US navy, raytheon demo network-enabled tomahawk cruise missiles in flight," *Microwave Journal*, vol. 58, no. 11, p. 48, 2015.

29. G. D. Rowley. (2017). Armed drones and targetted killing policy implications.

30. General Atomics Aeronautical. (2015). *Predator B: Persistent multi-mission ISR* [Online]. Available: www.ga-asi.com/Websites/gaasi/images/products/aircraft_ systems/pdf/Predator_B021915.pdf.

31. General Atomics Aeronautical. (2015). *Predator C avenger: Next-generation multimission ISR* [Online]. Available: www.ga-asi.com/Websites/gaasi/images/products/ aircraft_systems/pdf/Predator_C021915.pdf.

32. C. Enemark, *Armed Drones and the Ethics of War: Military Virtue in a post-heroic Age*. Taylor & Francis, 2013.

33. G. I. Seffers. (2017). NATO studying military IoT applications. *Signal*. Available: www.afcea.org/content/?q=Article-nato-studying-military-iot-applications.

34. H. Okcu, "Operational requirements of unmanned aircraft systems data lnk and communication systems," *Journal of Advances in Computer Networks*, vol. 4, no. 1, pp. 28–32, 2016.

35. G. Allen and T. Chan. (2017). *Artificial intelligence and national security* [Online]. Available: www.belfercenter.org/sites/default/files/files/publication/AI%20 NatSec%20-%20final.pdf.

36. H. U. Zaman *et al.*, "Design, control & performance analysis of Muktibot," *Presented at the 2016 IEEE 7th Annual Information Technology, Electronics and Mobile Communication Conference (IEMCON)*, Vancouver, B.C., Canada, 2016.

37. P. Paganini. (2016). *NATO officially recognizes cyberspace a warfare domain* [Online]. Available: http://securityaffairs.co/wordpress/48484/cyber-warfare-2/ nato-cyberspace-warfare-domain.html.

38. US Department of Homeland Security, "Strategic principles for securing the Internet of Things (IoT)," p. 17, 2016.
39. C. Brodie. (2008). *The importance of security awareness training* [Online]. Available: www.sans.org/reading-room/whitepapers/awareness/importance-security-awareness-training-33013.
40. A. Baldwin, I. Gheyas, C. Ioannidis, D. Pym, and J. Williams, "Contagion in cyber security attacks," *Journal of the Operational Research Society*, vol. 68, no. 7, pp. 780–791, July 01, 2017.
41. I. Nikolov. (2017). *Are cross-protocol attacks the next big cybersecurity danger?* [Online]. Available: www.forbes.com/sites/forbestechcouncil/2017/05/08/are-cross-protocol-attacks-the-next-big-cybersecurity-danger/#2f530d772fc0.
42. C. Paar *et al.*, "DROWN: Breaking TLS using SSLv2," in *25th USENIX Security Symposium*, Austin, TX, 2016: USENIX Association.
43. S. K. Rath, "South Asia's cyber insecurity: A tale of impending doom," *Qatar Foundation Annual Research Conference Proceedings*, vol. 2016, no. 1, p. ICTPP1054, 2016/03/01.
44. R. Sabillon, J. Cano, V. Cavaller, and J. Serra, "Cybercrime and cybercriminals: A comprehensive study," *International Journal of Computer Networks and Communications Security*, vol. 4, no. 6, pp. 165–176, 2016.
45. S. R. B. Prahbhu, M. Pradeep, and E. Gajendran, "Military applications of wireless sensor network system," *A Multidisciplinary Journal of Scientific Research & Education*, vol. 2, no. 12, 2016.
46. D. D. Coleman, D. A. Westcott, B. E. Harkins, and S. M. Jackman, *CWSP Certified Wireless Security Professional Official Study Guide*, 1st ed. Boston, MA: Sybex, 2010.
47. T. McGuiness. (2001). *Defense in depth* [Online]. Available: www.sans.org/reading-room/whitepapers/basics/defense-in-depth-525.
48. OWASP. (2015). *Defense in depth* [Online]. Available: www.owasp.org/index.php/Defense_in_depth.
49. W. Wade. (2015). *CIS updates the 20 critical security controls* [Online]. Available: www.tenable.com/blog/cis-updates-the-20-critical-security-controls.
50. SANS, "CIS Critical Security Controls poster," 41st ed., 2016.
51. Z. Hawkins. (2016). *Digital land power: The Australian Army's cyber future* [Online]. Available: www.aspistrategist.org.au/digital-land-power-australian-armys-cyber-future/.
52. Australian Government Department of the Prime Minister and Cabinet, "Academic cyber centres to tackle critical skills shortage," June 14, 2017.
53. Australian Government Department of Industry Innovation and Science, "$50 million investment into cyber security research and industry solutions," September 22, 2017.

Index

Printed in the United States
by Baker & Taylor Publisher Services